Prehistoric Human-Environment Interactions: People, Fire, Climate, and Vegetation on the Columbia Plateau, USA

Elizabeth A. Scharf

BAR International Series 2006
2009

Published in 2016 by
BAR Publishing, Oxford

BAR International Series 2006

Prehistoric Human–Environment Interactions

ISBN 978 1 4073 0582 0

BAR Publishing is the trading name of British Archaeological Reports (Oxford) Ltd.
British Archaeological Reports was first incorporated in 1974 to publish the BAR
Series, International and British. In 1992 Hadrian Books Ltd became part of the BAR
group. This volume was originally published by Archaeopress in conjunction with
British Archaeological Reports (Oxford) Ltd / Hadrian Books Ltd, the Series principal
publisher, in 2009. This present volume is published by BAR Publishing, 2016.

Printed in England

BAR
PUBLISHING

BAR titles are available from:

BAR Publishing
122 Banbury Rd, Oxford, OX2 7BP, UK
EMAIL   info@barpublishing.com
PHONE   +44 (0)1865 310431
FAX   +44 (0)1865 316916
www.barpublishing.com

Abstract

Modern ecological studies are unable to examine long-term processes operating on the order of hundreds of years. Because of the limited length of modern and historic records, questions about long-term interactions between people and the environment can only be answered using paleoecological and archaeological information. This volume presents prehistoric records that span over a millennium to examine issues of human paleoecology on the Columbia Plateau of Washington State, USA. Unlike many previous studies, this study (1) quantifies past human population, (2) compares relative inputs of humans, climate, fire, and vegetation using multivariate statistics, (3) examines relationships between variables when leads and lags of different lengths are introduced, and (4) identifies multicollinearity, allowing variables of no unique explanatory value to be eliminated.

For this analysis, lake sediments provide pollen, charcoal, and oxygen-isotopes that serve as proxies for past vegetation, fire, and climate, respectively. These data are compared to a previously published paleodemographic reconstruction. Results show that simple bivariate analyses are misleading. For example, human population is significantly correlated with charcoal when charcoal is lagged by 0, 50, 100, or 150 years, indicating that human action has led to future increases in fire. When all variables are entered simultaneously in a multivariate analysis, however, population has no statistically significant effect on fire after 50 years. Vegetation has had a significant short-term influence on fire, but climate is identified the only long-run predictor of fire. Similar analyses run to determine which factors (if any) were patterning future values of vegetation, human population, and climate show that the history of the system had an important influence on ecological outcomes. Analyses demonstrate that the different factors in the ecosystem are intricately interrelated to one another, with feedbacks occurring between the variables over several different time scales.

This study indicates that research on human impacts that focuses on bivariate patterns, such as simple comparisons of coeval human population and fire, can suffer from the problem of equifinality. The multivariate statistical procedures employed in this work avoid these problems, however, and can be used in any study that employs observations taken at equally-spaced time intervals. Additionally, the protocols developed and used in this volume can be easily adapted and applied in new geographical areas—the methods and research design used need not be tied to this particular location.

Table of Contents

List of Figures

List of Tables

Acknowledgements

Many people contributed to the production and refinement of this research project, and the resulting volume, which started as the author's dissertation project. Aid in developing a research strategy was given by Drs. Donald Grayson, Robert Dunnell, Matsuo Tsukada, Eric Smith, and Julie Stein, all of whom donated their time to help with reviewing and editing earlier versions of this work. Kathy Mauz, Martin Hoyer, and Drs. Larkin Hood, Kristine Bovy, Susan Hughes and Robert Kopperl were all indispensible in the field-work stage of the process. Field and laboratory work were supported by funding from the National Science Foundation (NSF award # SBR-9700544). Robert Lee and Zane Johnson provided technical aid and support in the production of the text and graphics. The University of North Dakota provided resources and facilities needed for preparation of this manuscript. Thanks also go to Dr. Anthony Cagle, who gave guidance, advice, and editorial feedback for this BAR volume.

Chapter 1. Introduction: Environment-Human Interactions

This volume presents a protocol that creates a set of metric proxy records for human population, vegetation, fire, and climate sampled at fifty-year intervals stretching back a little over a millennium. This resulting data set can be used to quantitatively assess the relative contributions and interactions of these variables at one point in time, or when 50, 100, 150, or even 300 year lags are introduced into the statistical model. Results from the analysis add to our knowledge of long-term ecological processes by examining the greater historical context of one particular location—Okanogan and Douglas counties in the state of Washington, USA. Such a study helps to delineate and define the role of prehistoric hunter-gatherer groups in their local ecosystems, even though this group is low in population density and practices foraging, gathering, and fishing, leaving more subtle traces than more populous farming groups. The results provide information on the significance and extent of interaction between different factors, including anthropogenic forces, in the ecological system. In this way, historical influences, complex feedback systems, and independent causal factors can be identified and explored to better understand the past, present, and future of the environment in eastern Washington. Both the methods used and the study-specific results gained from this research contribute to the larger goals of documenting and understanding long-term landscape processes around the world. The research design presented here can be applied at other times and places on the globe, providing a systematic means of evaluating the contribution of archaeological groups to the structure and function of plant communities.

Interest in Human-Environment Interactions

How people and the environment interact is an object of continued concern and interest, worldwide (Head 2007). One need only to access the web, pick up a newspaper, or turn on a radio or television to encounter headline news about such issues as global warming, the destruction of wetlands, deforestation, extinction, El Niño, and La Niña. Human-environment interaction is of great interest to us for two reasons. First, as humans, we are innately interested in the legacy that we will leave behind in the form of our impact on the landscape and ecology of the planet. We want to understand how we affect the environment and how to be more knowledgeable stewards of our resources (Black 1970, Goudie 2005). Underlying this is the second and more important reason for our interest—we are concerned that changes in the climate, flora, or fauna may have a harmful impact on our future as individuals and as a species.

Why Prehistory Matters

There are many stake-holders with an interest in how current events are shaping our ecological futures. The problem is that many climatological and ecological responses take longer than a few years, or even a human lifetime, to manifest themselves. Vegetation responses to present-day disturbances, for example, could take many decades to recognize. Plant succession takes a hundred years or more to reach a state of "climax," so the result of present perturbations will take over a hundred years to be fully realized (Birks 1981, Delcourt and Delcourt 1991, Horn 1975). Unfortunately, most ecologists "study organisms on a time scale of less than 100 years (and are usually limited to the time span of a field season, grant or dissertation)" (Schoonmaker and Foster 1991:205).

To better understand the full ramifications of slowly operating processes, a time-frame is needed that is longer than modern ecology can deliver. Greater time-depth is exactly what the disciplines of paleoecology and archaeology have to offer. As many have observed (to the point of it becoming cliché), knowledge of the past is the key to understanding both the present and the future. In recognition of this, many researchers have used paleorecords to put ecological questions into a larger context and help guide management decisions (Arno and Delcourt and Delcourt 1991, Gavin et al. 2007, Sneck 1977, Schoonmaker and Foster 1991, Swetnam et al. 1999, Wray and Anderson 2003).

And yet, succession and other ecological processes do not function the same way in different contexts. Contemporary starting conditions are important but so is the history, the prior conditions of a system (Foster 2000). Using modern observations and information on current environmental conditions and past cultural use, ecologists have surprisingly established that human action has a more important and lasting effect on vegetation than previously realized. Working in New England, Gerhardt and Foster (2002) demonstrated that land use from 18th and 19th centuries was more significantly and tightly correlated with vegetation than any other explanatory factor (such as slope, aspect, elevation, soil texture, or recent damage from fires or hurricanes). Obviously, a long-term view that can accommodate individualistic trajectories of change with potential anthropogenic input is required for understanding ecological functions of the past, present, and future.

Research Questions

The project presented here seeks to add to this growing body of knowledge using a Holocene dataset to understand long-term processes and outcomes. This research aims to explicate the interactions between humans, climate, and the landscape over the past millennium using paleoecological, paleoclimatological, and archaeological records. Specific questions examined along the way include: How do humans influence fire frequency and how does that, in turn, affect humans? How does climate control fire frequency? How does fire frequency shape vegetation? How does vegetation impact humans? How does fire frequency affect humans? How does climate have an effect on human populations? What kind of constraints does the history of a system impose? And how do all these factors interact with one another in more complex ways?

Human-Environment Interactions in Prehistory

Questions about human- environment interactions, such as those posed above, are not new. Studies of human-environment interactions abound, whether these are based on contemporary observations, written and oral histories, or reconstructions of prehistoric processes. Of these, truly long-term (decadal-, centennial-, or millennial-scale) processes in many parts of the world can only be examined using prehistoric databases that predate the advent of written records. Indeed, prehistoric records have been used to address a wide variety of issues, including the impact of climatic and ecosystem changes on human population sizes and subsistence practices, as well as the influence of agricultural land clearance on vegetation and soil erosion, the effect of human colonization on fire regimes, the impact of human hunting on wild animals, and the ramifications of introducing domestic livestock into new areas (Redman 1999, Redman et al. 2004).

How do such studies identify changes in cultural practices caused by environmental changes? How are human impacts isolated from other confounding factors? How are the effects of humans recognized, since prehistoric human action cannot be directly observed but must be indirectly inferred from records of past environmental conditions?

Research on Prehistoric Human Impacts

Assessments of human impact on a given biota primarily rely on changes in the type, number, and relative abundances of plant and animal taxa in the past. Human impacts on the faunas of such geographically distant regions as western North America (Broughton 1994, 1995; Grayson 2001; Grayson and Cannon 1999) and New Zealand (Nagaoka 2000, 2001), for example, have been studied employing the diet breadth model provided by foraging theory. Decreasing relative abundances of high-ranking (typically large-bodied) prey in archaeofaunas over time are attributed to human impact (Grayson 2001, Grayson and Cannon 1999). Release of species from predation pressure and competition with people has also been suggested for the Protohistoric era of North America. Declining Native American populations have been cited as the reason for possible expansions in the spatial distribution of bison and elk (Kay 1994, Mann 2005), and for the greatly increased numbers of bison, deer, and elk (Broughton 1995, Mann 2005) seen by later Europeans and Euro-Americans.

Arguments have also been made for human impacts in cases of extinction, in which researchers argue that decreasing numbers of animal taxa following human colonization of new areas are evidence of human impact (Redman 1999). These studies range from the relatively

secure documentation of anthropogenic extinction of birds on small Polynesian islands (Steadman 1995, Steadman and Kirch 1990) following human colonization, to the more highly debated extinction of Pleistocene megafauna that some attribute to the human settlement of areas like North and South America (e.g., Grayson 1991, 2001; Martin 1967, 1972, 1984; Martin and Steadman 1999).

Anthropogenic Vegetation Changes

Assessments of human impact on past vegetation, likewise, focus on taxonomic representation in the prehistoric record. Assessments of anthropogenic change in vegetation are often based on evidence from pollen diagrams. One clear indicator of anthropogenic change, recognized worldwide, is the sudden appearance of exotic plant taxa in new geographic areas (Cole and Liu 1994, Delcourt and Delcourt 1987b). The introduction of weeds, such as common plantain (*Plantago major*) and dandelion (*Taraxacum officionale*) in North American sediments after European Contact, is an example of one of these exotic "markers" of human action (McAndrews 1976, 1988). The appearance of cultigens and domesticated plant taxa into new areas is used as an indicator of human impact as well. The deposition of maize (*Zea mays*) pollen in North America (Delcourt and Delcourt 1987a, 1987b), the spread of cassava (*Manihot esculenta*) pollen into new parts of the Amazon Basin in South America (Behling and da Costa 2000), the introduction of domesticated cereals (*Cerealea*) into Europe (Iverson 1956) and rises in buckwheat (*Fagopyrum*—Tsukada et al. 1986) during the Jamon period in Japan are just a few examples of such indicators.

Human disturbance also favors taxa that are typical of early-successional communities over those that are characteristic of late-successional communities. This is commonly expressed on the landscape by the conversion of woodlands into more open vegetation, such as grasslands. This is documented in pollen records by the overall increase in herbaceous taxa at the expense of arboreal taxa in pollen spectra over time. When such disturbance is temporally correlated with human behavior, in the absence of other potential causal mechanisms (such as climate change), the resulting vegetation changes are identified as anthropogenic (Abrams and Nowacki 2008, Baker 2002). Changes in vegetation that are patchy in time and space are also considered evidence of human action as humans are assumed to create small-scale localized changes relative to larger-scale climatic influences (Janssen 1986).

Investigations at Black and Tuskegee Ponds in Kentucky used these criteria to document anthropogenic influences over a 9,400 year time frame (Chapman et al. 1989). Data from these sites revealed that human clearance began with Native American populations during the Archaic period, and accelerated around 1,500 years ago in the uplands (Delcourt and Delcourt 2004, Delcourt et al. 1986). In a similar approach, Delcourt and others (1998) examined 32 samples from a 9,500 year pollen and charcoal record at Cliff Palace Pond Kentucky. They determined that people have been using fire on this continent for thousands of years, resulting in an anthropogenic environment with more spatially patchy vegetation that favored more fire-tolerant oaks, chestnuts, and pine over climax vegetation in upland forests (Delcourt and Delcourt 2004, Delcourt et al. 1998). They also noted a 3,000-year record of increased disturbance indicators such as bracken fern, ragweed, and cultigens such as sunflower, sumpweed (marshelder), and goosefoot due to agricultural and horticultural activities of Native Americans (Delcourt et al. 1998). Work in North Carolina at Horse Cove, using pollen and charcoal counts from pollen slides, has also demonstrated the prehistoric prevalence of anthropogenic fire and the role of humans in creating and maintaining the vegetation that existed on this continent before Contact (Delcourt and Delcourt 1998).

This pattern was found to hold for all the deciduous forests of eastern and southern North America as a whole. During the late Holocene these forests were dominated by shade-intolerant species such as oak, hickory, and chestnut with little understory, interspersed with open patches. The abundance of these taxa over the past four thousand years has been linked to Native American activities that favored a mosaic of early- to mid-successional taxa (Hammet 1992). Native people discouraged climax taxa through transplanting, use of frequent low-intensity fires, clearance activities, and girdling of unwanted species. These activities promoted a patchwork of prairie, savanna, and open oak-dominated forests (Abrams and Nowacki 2008). After the introduction of European diseases to the New World after A.D. 1500, the depopulation of native peoples resulted in decreased anthropogenic disturbances in this region. As a result, the eastern woodlands of North America have become more homogeneous, with closed canopies, favoring late-successional shade-tolerant trees such as maple, hemlock, and white pine over the early- to mid-successional oaks and other taxa that formerly dominated the area (Abrams 2005, Abrams and Nowacki 2008, Black et al. 2006, Buckner 2000, Delcourt et al. 1998, Dorney and Dorney 1989, Hicks 2000, Kay 2000, Mann 2000). Similar cultural activities have also been credited with creating and maintaining savannas, barrens, and prairies elsewhere (outside of wooded areas) in the central and eastern United States as well (Abrams and Nowaki 2008, Heikens and Robertson 1994).

Other clearance episodes have been documented in prehistoric contexts for cultures around the world. New Zealand, for instance, was heavily forested until the arrival of the Maori. Following the Polynesian settlement of New Zealand, arboreal taxa (e.g., podocarps) decreased radically while disturbance taxa (e.g., grasses and ferns) increased in relative abundance on the landscape (McGlone 1983, McGlone and Wilmshurst 1999). Like New Zealand, Easter Island, Mangaia (Kirch et al. 1992), and the Hawaiian Islands also suffered from deforestation following Polynesian settlement. In the case of Easter Island, settlement of the island around A.D. 400 was followed by decreases in palm trees and increases in

grasses, a trend that continued to the local extinction of the palms around A.D. 1400 (Diamond 1995, 2005; Flenley et al. 1991; Flenley and King 1984—see Hunt and Lipo 2006 for a different view). Similar processes operated on Hawaii, which was also settled around A.D. 400. As the abundance of palm trees decreased sharply over time, disturbance taxa such as grasses and ferns increased (Athens and Ward 1993).

Patterns of anthropogenic change like those documented during Polynesian settlement of the Pacific are typical of prehistoric human impact seen across the globe. Large-scale disturbance has been documented for a wide range of locations and cultures, from the deforestation of the classic Maya period in the Yucatan Peninsula (Abrams and Rue 1988, Mann 2005), the Neolithic clearance of lands in the British Isles (Brown 1997, 1999; Greig 1992; Macklin et al. 2000; Skinner and Brown 1999) and Euro-American clearance in western Washington (Sugita and Tsukada 1982) to the drastic reduction of woodlands in the Anyang Dynasty in China (Redman 1999). In prehistoric North America, the fall of Cahokia has been linked to deforestation and logging (Mann 2005). In the American Southwest, the people of Chaco Canyon had such a great need for timber that they transported logs over 75 km to their pueblos (Allen 2002). Many researchers in the area conclude that the forests near Chaco Canyon have still not recovered from the overharvesting of trees that occurred over 800 years ago (Swetnam 1999).

Human disturbance is often associated with increases in both fire frequency and soil erosion, and these usually coincide with decreases in late-successional taxa. In the case of the Maya, deforestation was accompanied by both fires and massive erosion, as evidenced by the increased deposition of both charcoal and inorganic sediments into area lakes and oceans (Binford et al. 1987, Hodell et al. 2005, Mann 2005, Rice 1996, Rice and Rice 1984, Rice et al. 1985). Likewise, human clearance of palms in Hawaii was correlated with an increase in anthropogenic fire and erosion (Kirch 1982, 1997; Kirch and Hunt 1997). Deforestation on Easter Island (Diamond 1995, Flenley and King 1984, Flenley et al. 1991) and China (Redman 1999) both resulted in soil loss, and the loss of forest cover following the settlement of New Zealand was accomplished through human-set fires (McGlone 1989, McGlone and Wilmshurst 1999, Ogden et al. 1998).

Anthropogenic Fire

Fire is credited as one of the most universal and effective causes of human-induced environmental change (Stewart 1956). As with the case of anthropogenic vegetation disturbance, human influence on fire is often recognized by first identifying an increase in charcoal deposition that is not correlated with known changes in the climate or other external environmental factors. If this otherwise unexplained increase in fire can be shown to be correlated in time with a significant archaeological event such as initial settlement of an area, adoption of agriculture, or intensified land-use from rising population densities, then the rise in charcoal is deemed evidence of anthropogenic fire (Whitlock and Knox 2002). This method of identifying episodes of anthropogenic firing has been used to identify prehistoric human impacts in such far-flung areas as New Zealand (McGlone 1989, McGlone and Wilmshurst 1999), the Falkland Islands (Buckland and Edwards 1998), the Amazon Basin (Behling and da Costa 2000, Mann 2005), Australia (D'Costa et al. 1993; Head 1989; Jones 1969; Pyne 1991, 1997; Singh 1980), Hawaii (Kirch 1982), the Faroe Islands (Hannon and Bradshaw 2000), Chile (Moreno 2000), Madagascar (Burney 1993), the Yucatan (Mann 2005, Rice 1996, Rice and Rice 1984), South India (Morrison 1994), the Pacific Northwest (Brown and Hebda 2002a, 2002b; McDadi and Hebda 2008), the Great Plains (Abrams and Nowacki 2008), the southern and eastern deciduous forests of North America (Albert 2007; Black et al. 2006; Chapman et al. 1989; Clark and Royall 1995a, 1996; Delcourt 1987; Delcourt et al. 1998; Delcourt and Delcourt 1985, 1998, 2004; Patterson and Sassaman 1988; Whitehead and Sheehan 1985), and possibly California and the Sierra Nevadas (Aschmann 1977, Bendix 2002, Keeley 2002).

Besides causing an increase in charcoal influx, anthropogenic fire is assumed to result in a different season, intensity, frequency, spatial location, interannual variability, and spatial homogeneity of burns (Allen 2002; Griffin 2002; Kay 2000, 2007; Lewis 1982, 1993). Fire scar histories record the season of past burns. In many parts of the west, lighting fires occur in summer, yet tree ring information indicates that many prehistoric fires burned in winter, spring, or fall—times when many researchers feel that people were the dominant ignition source (Kay 2000). For example, Kay (2000) points out that "asbestos" aspen communities in the Rocky Mountains burn infrequently in modern times because lightning strikes occur only in summer when the trees are green and make poor fuel. In the past, aspen forests burned frequently because they were ignited when trees were leafless and the undergrowth was dry and flammable. Kay (2000) argues that the fires started outside the summer season in stands of aspen because native people set them. Along with others, he likewise reasons that eastern forests of North America were also subject to aboriginal burning because they will only burn well when the trees are leafless, which happens in seasons when lightning strikes are few and ineffectual (Buckner 2000, Kay 2000). Increased numbers of low intensity burns are also credited to human action. For example, accelerated burning in the Jemez and other mountains in northern New Mexico in the American Southwest are credited to prehistoric cultural activities (Allen 2002). In the Great Basin, areas in the northern part of the region experienced more frequent fires because they were more densely populated and contained more travel corridors (Griffin 2002). These fires were more constant from year to year than lightning ignitions, as well (Griffin 2002). Spatial heterogeneity is often cited as a characteristic of anthropogenic fire (Allen 2002, Vale 2002a). Impacts are assumed to be highest near areas frequently used or inhabited by humans (Baker 2002; Black et al. 2006; Vale 2002a, 2002b). Although

some feel that climate and lightning fires are adequate for explaining most fires in the American west (e.g., Baker 2002, Bendix 2002, Millspaugh et al. 2000, Mohr et al. 2000, Parker 2002b, Vale 2002a, Whitlock and Knox 2002), Kay (2007) compiled and analyzed data on lightning strikes and burns in national forests in the United States, and showed that lightning mathematically accounts for a minority of burns. His research confirmed that anthropogenic fire is needed to explain the high observed fire frequency from pre-Columbian times. Kay (2007) also calculated that even with a low population density and low rate of ignitions, human sources accounted for more fires than natural causes.

Quantitative comparisons evaluating the statistical significance of natural versus cultural factors come from only two locations in North America. Guyette and others (Guyette and Dey 2000; Guyette and Spetich 2003; Guyette et al. 2002, 2006) compiled measures of Native American population during the historic era from census records in the Boston Mountains of Arkansas. Population densities, topographic roughness, fire scar records, and climate were all compared using multivariate regression. Results showed that at low population levels, the single best predictor of fire was the number of native people present on the landscape (Guyette and Dey 2000; Guyette and Spetich 2003; Guyette et al. 2002, 2006). A study from the Allegheny Plateau in Pennsylvania quantified prehistoric human presence by looking at the kinds and distribution of archaeological sites in the area. Black and others (2006) created an index of Native American influence (NAI) that varied spatially, with areas of greater intensity near known villages and travel routes (Black et al. 2006). Linear regression showed that NAI was the most significant predictor of the distribution of early successional trees (oak/hickory/chestnut) than were natural factors such as elevation, slope, aspect, landform, and substrate (Black et al. 2006).

Requirements for a Study of Prehistoric Human Impact

Examining these existing studies of anthropogenic impacts helps identify the requirements that must be met when designing a new research project. (1) Proxies must be available for the past biota, human behavior, fire, and climate. (2) These proxies must be constructed on a temporal and spatial scale compatible with the scale of human action. (3) There has to be adequate temporal control for the surrogates, enabling them to be meaningfully compared. (4) There has to be a change in state of at least one of the variables. (5) The records being used must have enough time-depth to address the questions being asked.

Additionally, another criterion must be met for this study. Although past studies have identified human impact, this study seeks to examine and quantify these interrelationships on a finer scale than most. Relationships between contemporaneous variables are investigated by calculating correlation coefficients (Pearson's *r*). Correlations are also run with lags introduced between variables to identify which variables are leading (likely to be causal) and which are lagging (likely to be responding variables). Further, time series (in the time domain) and Granger causality are used to simultaneously evaluate the relative contribution of all factors (human action, climate, fire, vegetation), both contemporaneous and leading, to changes in the system. In this way, Granger causality can identify which factors (human action, climate, fire, vegetation), if any, are driving the system. It allows for feedbacks, dependencies, and external factors to be explored quantitatively. Because of this, problems of equifinality can be addressed. Accordingly, for this study, surrogates need to be expressed numerically so the nature and strength of the relationships between them can be examined statistically. This study contrasts with earlier studies of prehistoric human impact, in that it uses multivariate statistics to simultaneously evaluate multiple competing explanatory variables relative to one another on multiple time scales.

Meeting the Requirements

The research in this volume has been designed to meet all of the above criteria. First, the chosen study area (the area immediately surrounding Rufus Woods Lake in eastern Washington) is an appropriate size, and has an excellent pre-existing population proxy. The paleodemographic record for this area extends back over a millennium, is sampled at evenly-spaced and well dated 50-year intervals, and exists in a highly useful quantitative form (Campbell 1989, 1990).

Since no comparable surrogates for vegetation, fire, or climate could be found in the literature for this particular study area, new records for these variables had to be specially created for the current study. Information on past plant communities is drawn from fossil pollen, a common paleovegetation proxy. Since permanently wet lakes preserve continuous palynological records in excellent condition, lake sediments provided the data source for reconstructing paleovegetation. Fortuitously, lake sediments can also provide information on fire and climate as well. Climatic information is drawn from the isotopic composition of lake carbonates, and charcoal influx is used to track past vegetation fires. Finally, chronological control is provided by a series of AMS radiocarbon dates. The rest of this volume gives details on the region, study area, and research design. Results of AMS dating, isotope assays, pollen analyses, and charcoal counts are presented and interpreted separately, and then analyzed simultaneously. Statistical analyses and a synthesis of results allow for the greater implications that are drawn and presented at the end of this report.

Chapter 2. Selecting a Study Region: The Plateau Culture Area of North America

Archaeological Requirements

Dependable, systematic data regarding past human behavior are critical to any investigation of human involvement in paleoecological systems (Black et al. 2006). To be successful and useful, a study of human-environment interaction must include an estimation of the number of people living in a given area in order to evaluate their relative influence on the landscape compared to other factors, such as climate and topography (Guyette and Dey 2000; Guyette and Spetich 2003; Guyette et al. 2002, 2006). For times that predate written records, this information can only be obtained indirectly, through a proxy demographic record. Although some researchers have relied on guesses, anecdotal information, or extrapolations from assumed ecological carrying capacity or historical records to estimate past human population sizes (Denevan 1996), the use of these lines of evidence have not led to a resolution to the debate over the timing, extent, and magnitude of changes in Native American populations (Campbell 1989, 1990; Denevan 1996; Kay 2000; Mann 2005; Thornton 1997; Vale 2002a). Instead, it has been argued that good proxy demographic records can only be produced with archaeological data as surrogates (Black et al. 2006; Campbell 1989, 1990; Cook 1976; Dobyns 1963, 1983; Dunnell 1991; Ramenofsky 1987), compiled separately for each local group (Kealhofer and Baker 1996).

Prehistoric human populations have been reconstructed using many lines of information obtainable from the archaeological record. Past population estimates have been based on measurable archaeological phenomena such as burial counts, archaeological site abundances, house counts, hearth or storage-pit tallies, the number of radiocarbon dates, wear on grinding stones, volume of pottery vessels, wear on floor tiles, floor area inside buildings, total site area, midden volume, room totals, proximity of known archaeological sites, or the amount of accumulated shell and bone midden (e.g., Black et al. 2006; Campbell 1989, 1990; Casselberry 1974; Cook and Treganza 1950; Naroll 1962; Plog 1975; Ramenofsky 1987; Schacht 1981; Schwartz 1956; Shawcross 1967, 1972; Turner and Lofgren 1966). All of these suggested measures rely on the assumption that the amplitude of human impact on the archaeological record is proportional to the number of people on the landscape, with greater numbers of people producing greater amounts of food waste, creating more wear on tools, or constructing more living space for themselves. Generating a secure, reliable, and fine-grained human population reconstruction using these kinds of data is an undertaking that requires the excavation, laboratory analysis, and radiocarbon-dating of hundreds of site components in a small geographic area. Such projects usually require years of effort to complete, and become the subject of books or dissertations by themselves. To construct a new record for an area would be prohibitively costly in terms of time, labor and money. Such a large archaeological undertaking could also involve destructive analysis, making the use of existing data sets that much more desirable, both logistically and ethically. For these reasons, an area with an existing systematic prehistoric population proxy was sought for this study.

For optimal utility, a paleodemographic record should be constructed on a spatial and temporal scale consistent with human behaviors (Stein 1993). The scale of the population surrogate must be large enough to encompass a group of people over several generations, yet fine-grained enough to be relevant to the life and actions of a single person. The population record must match the scale of human disturbance in the paleoecological record. An appropriate temporal and spatial scale for such an investigation is what authors alternatively refer to as mesoscale (Delcourt et al. 1983, Delcourt and Delcourt 1988, Schoonmaker and Foster 1991) or extralocal (Jacobson and Bradshaw 1980). This scale encompasses areas from approximately tens to a thousand square kilometers, but tending to be on the order of 100 square kilometers. Mesoscales also span time periods of several decades to a millennium or two. Although this determines the duration of time that needs to be covered, the temporal scale also determines the necessary resolution of the proxy; the time-span between data points must be decadal to centennial in scale. Areas with sufficient resolution and a good existing proxy demographic record covering a millennium are quite rare. This makes the proxy demographic record the limiting and essential factor in selecting a research area and key to successfully evaluating the effects of prehistoric populations on the landscape.

There are a handful of regions in North America for which such diachronic population reconstructions have been produced (e.g., Ames 2000; Campbell 1989, 1990; Chatters 1995a, 1995b; Ramenofsky 1987). These include such areas as upstate New York (Ramenofsky 1987), the Caddo region (Perttula 1992), the Middle Missouri River Valley (Ramenofsky 1987), the Lower Mississippi River Valley (Ramenofsky 1987), and the Southern Plateau (Ames 2000; Campbell 1989, 1990; Chatters 1995b).

Of these areas, the Southern Plateau was selected for study. One advantage of the Southern Plateau was that it had the potential to provide a record with tighter chronological control than other regions. The Southern Plateau has frequently been covered by easily identifiable volcanic ashes of known composition and known ages, providing secure and abundant time markers throughout the Holocene record (Fiacco et al. 1993; Hoblitt, et al. 1980; Sarna-Wojcicki and Davis 1991; Sarna-Wojcicki et al. 1991; Smith et al. 1979; Westgate and Gorton 1981; Yamaguchi 1983, 1985). In contrast, there is no similar series of well-dated stratigraphic events that could provide chronological control over the entire Caddo region, Lower Mississippi River Valley, upstate New York, or Middle Missouri River Valley. A second advantage of the Plateau is that it provides an opportunity to investigate prehistoric North American hunter-gatherer impacts on vegetation and the pollen record, and the effects of depopulation in areas used by hunter-gatherers, which are two issues that have gotten little treatment in the literature and deserve more attention (Kay 2000, Lacourse et al. 2007, McAndrews 1988). Most studies of the impact of European contact on Native American populations have focused on farming populations and it has been assumed that demographic disruptions had a much greater impact on socio-politically complex groups than on hunter-gatherer populations (Kealhofer and Baker 1996). Likewise, previous palynological attempts to document prehistoric impacts in North America have concentrated on high-density areas and horticultural populations (e.g., Albert 2007; Black et al. 2006; Chapman et al. 1989; Clark and Royall 1995a, 1996; Delcourt 1987; Delcourt et al. 1998; Delcourt and Delcourt 1985, 1998, 2004; Patterson and Sassaman 1988; Whitehead and Sheehan 1985). Also, paleoecological research is a welcome addition to established work in this area, following a long tradition of modern fire ecology research, conservation efforts, and biodiversity projects (e.g., Martin et al. 1977).

The Southern Plateau Region

The Plateau is a relatively low-lying interior area east of the Cascade Mountains in the Pacific Northwest (Figure 2.1), and is geographically defined as that part of North America drained by the Columbia and Fraser Rivers (Walker 1998). The Southern Plateau is the part of the Plateau that lies south of the highest elevation in the Okanogan highlands, and can be defined as that portion of the Plateau that is drained by the Columbia River. The Southern Plateau, also called the Columbia Plateau, is considered to be a coherent geographic subdivision of the Plateau, both in terms of its environment and culture (Ames 2000).

Prehistoric Population

Population reconstructions for this entire region have been published by Ames (2000) and Chatters (1995b). Figure 2.2 shows these demographic surrogates, which are based on the number of radiocarbon dates reported from archaeological sites of different ages. In both cases, mathematical transformations were used to correct raw data for decay of organic materials, since older materials are more likely to have been destroyed and therefore unavailable for radiocarbon dating. This method assumes that the number of radiocarbon dates analyzed is related to the amount of charred material produced per time period and related to the number of people producing charred material in each time period (Ames 2000, Chatters 1995b). This, of course, assumes that archaeologists are discovering, sampling, and dating materials in an unbiased manner.

Unfortunately, these surrogates are drawn from a very large area, some 100,000 square kilometers in size, well beyond the optimal spatial scale suggested for investigating human impacts (Delcourt and Delcourt 1988). Thankfully, a smaller-scale proxy demographic reconstruction exists from a portion of the Plateau. The Rufus Woods Lake area of eastern Washington (Figure 2.3) was the focus of a population study by Campbell (1989, 1990), that produced a demographic proxy of the appropriate temporal and spatial scale. Rufus Woods Lake is a reservoir of the impounded Columbia River, stretching from the Chief Joseph Dam to the Grand Coulee Dam, and the study area includes the areas adjacent to it on the north (in Okanogan County) and to the south (Douglas County).

Use of Fire by North American Hunter-Gatherers

Historic records, oral histories, and ethnographies from the Southern Plateau and hunter-gatherer groups in neighboring regions indicate that people may have used fire to influence the nature of vegetation. Indeed, evidence of native firing is seen throughout western North America. Many researchers believe that anthropogenic fires on this continent determined the spatial distribution of the flora—creating and maintaining a mosaic of patches at different successional stages, increasing floral diversity, and encouraging the growth of economically important taxa (e.g., Day 1953, Denevan 1992, Dorney and Dorney 1989, Hammet 1992, Pyne 1982).

On the Great Plains, for example, routine native burning determined the areal extent of grassland prairie and both the size and spatial distribution of bison herds (Abrams and Nowacki 2008, Boyd 2002, Mann 2005). When routine firing of the grasslands ceased in historic times, grasses were soon overgrown by trees, converting the land to forest and decreasing bison forage (Goudsblom 1992, Lewis 1982). Historic eyewitness accounts of ethnographic burning on the Plains and the Rockies abound, dating back to the accounts of Lewis and Clark (Barrett and Arno 1999, Pyne 1982, Krech 1999). Likewise, Native Americans in Alberta and Manitoba set fires that helped maintain grasslands and a patchwork of microhabitats that encouraged the growth of economically important plant and animal resources, including berries, rabbits, moose, and geese (Lewis 1982, Lewis and Ferguson 1988).

Contact and post-Contact burning is also documented for hunter-gathers in California and Oregon. Photographic records from Yosemite show that after native groups abandoned the area in 1854, the cessation of regular burning led to an increase in brush and deciduous trees at the expense of meadows and open grassy clearings (Aschmann 1977). Scattered historic reports from missionaries and travelers indicate that indigenous peoples burned the California chaparral, keeping grassy areas open between stands of shrubs (Bean and Lawton 1992; Burcham 1974; Keeley 2002; Lewis 1982, 1993; Pyne 1982). Native Californians also set fires to kill mistletoe (Loranthaceae family) that infested mesquite

**Figure 2.1. Location of the Plateau Culture Area in North America
(Based on Walker 1998:iii)**

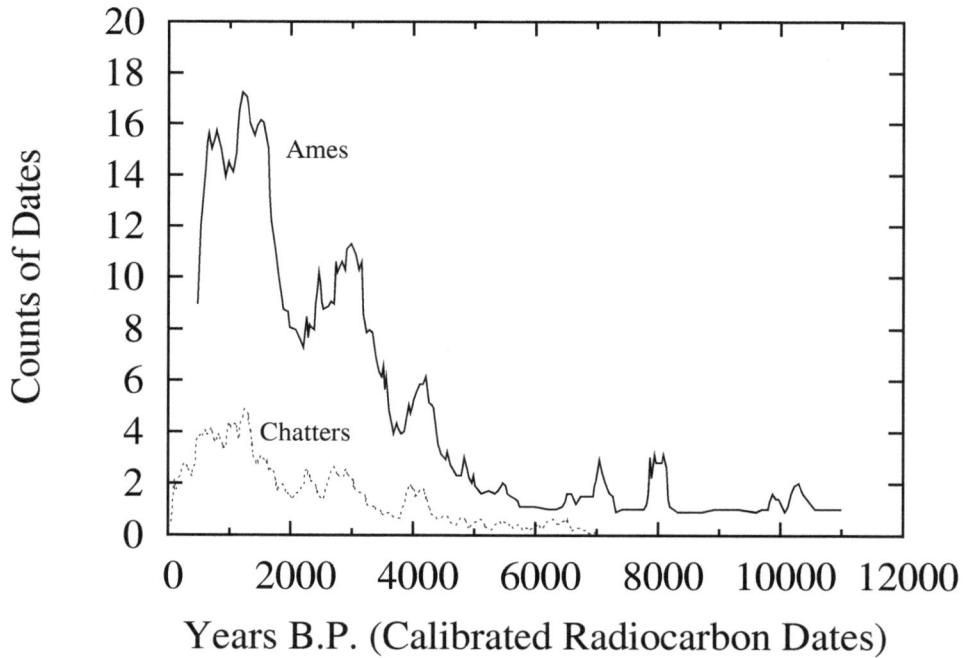

**Figure 2.2. Counts of Radiocarbon Dates on the Plateau as a Population Proxy
(Adapted from Ames 2000 and Chatters 1995b)**

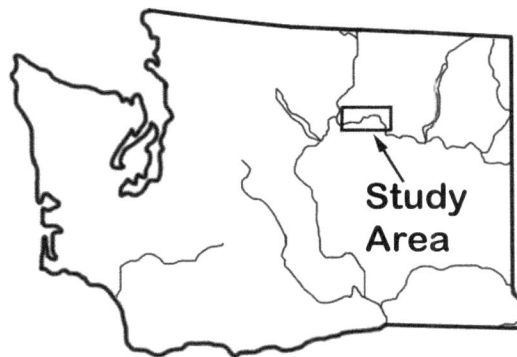

**Figure 2.3. Location of Study Area
(Rufus Woods Lake Area, Washington State)**

(*Prosopis*) and economically important stands of oak (*Quercus*) (Lewis 1993, Pyne 1982).

More complete ethnographic and historic records, combined with fire-scar histories of standing trees from forested areas in the Sierra Nevada and the Coast Range of California, indicate that native burning kept down undergrowth, dead wood, and forest-floor litter (Aschmann 1977, Keeley 2002). This, in turn, kept the coniferous forests open and prevented the kind of disastrous wildfires that destroy living stands (Lewis 1982, 1993). Established patterns and distributions of shrubs and grasses in this region cannot be explained by environmental factors such as climate and soil types, leading researchers concerned with modern land management such as Keeley (2002) to conclude that Native American firing practices are responsible for the ecosystems that we are currently trying to maintain or reconstruct (see Parker 2002b for a different interpretation of the evidence).

Throughout California and Oregon, grassland firing resulted in an artificially-induced late-fall sprouting of new grasses. This second growing season, in turn, provided winter forage that attracted deer and migratory birds (Bean and Lawton 1992, Lewis 1982, Pyne 1982, Timbrook et al. 1982). The evidence for anthropogenic environmental change in California is so abundant and so clearly documented that it has led some researchers to claim that "the vertical structure, spatial extent, and species composition of the various plant communities that early European visitors to California found so remarkably fecund were largely maintained and *regenerated* over time as a result of constant, purposive human intervention" (Blackburn and Anderson 1993:19).

It is no surprise, then, that native peoples in the Northwest also used fire as a management tool (Hunn 1990; Ignace 1998; Lewis 1973, 1977; Pyne 1982). Fire was used in the lowlands west of the Cascades by groups from Southern Oregon to British Columbia (Boyd 1999, Turner 1999). Low elevation burning was used to drive game and to promote the growth of plants that were

important to the diet of indigenous peoples. Ignition of vegetation promoted the growth of bracken fern (*Pteridium aquilinum*), acorns (*Quercus*), blackberries (*Rubus ursinus*), hazelnuts (*Corylus cornuta*), salmonberries (*Rubus spectabilis*), and other economically important resources (Boyd 1999, Deur 1999, Kruckeberg 1999, Turner 1999, White 1999). On the Olympic Peninsula, fire scar data, repeat photos, oral traditions, and historic documents indicate that native use of anthropogenic fire created and maintained prairies and a unique savannah ecosystem of bear grass (*Xerophyllum tenax*) and Douglas fir (*Pseudotsuga menziesii*) that disappeared when these cultural practices stopped (Peter and Shebitz 2006, Wray and Anderson 2003). Likewise, at the time of European Contact and settlement, many areas of Puget Sound and the Willamette Valley were open prairies or mosaics of woodlands and prairie openings that were anthropogenic in origin (Cox 1999, Kruckeberg 1991). After the imposition of historic fire suppression, late-successional vegetation took over these areas, with Douglas fir and conifers invading areas that were formerly kept open by Native American fires (Boyd 1999, Cox 1999, Kruckeberg 1991, Leopold and Boyd 1999). In British Columbia, fires during the late prehistoric era were attributed to native populations on southern Vancouver Island (Brown and Hebda 2002a, 2002b; McDadi and Hebda 2008). Increased charcoal influx coincided with archaeological evidence of significant human use of the area at a time when climate was becoming cooler and wetter—conditions not favorable for the outbreak of natural fires (Brown and Hebda 2002a, 2002b; McDadi and Hebda 2008). On Anthony Island, LaCourse and others (2007) found vegetation changes due to harvesting pressure put on economically useful plants by native peoples. Western red cedar (*Thuja plicata*) decreased during times when the Haida occupied the area, as they preferentially harvested red cedar for manufacturing purposes.

Accounts from oral histories and ethnographies show that Native American groups were using fire to manage resources at higher elevations in the Cascade Mountains as well. Sahaptin firing in what is now the Gifford

Pinchot National Forest continued in the Cascades until the recent imposition of fire suppression, and Indian burning is still remembered by living members of local communities (Hunn 1990). Vegetation burning by Sahaptin peoples in the Cascades was used primarily to enhance economically useful herbs and shrubs and to discourage trees (Hunn 1990, 1999). In this way, people encouraged the growth of berries (e.g., blueberry, grouseberry, and huckleberry) that were part of the human diet as well as the growth of vegetation that could feed deer and elk (animals which were also part of the human diet) (Hunn 1990, 1999). Photographic evidence from the Gifford Pinchot National Forest shows that past burning in this area kept meadows open and promoted the growth of early-seral taxa. Meadows that had been open and treeless in photos taken around A.D. 1900 were invaded during later decades of fire suppression, with young lodgepole pines (*Pinus contorta*) encroaching on the edges and spreading to the center of the clearings (Hunn 1990, Jerman and Mason 1976). Molala groups of the Cascades in Oregon were known to follow similar practices; these native groups also set fire to vegetation at higher elevations in order to increase both huckleberries and browse for elk and deer (Zenk and Rigsby 1998).

Fire was an equally important tool for the inhabitants of the intermontane areas east of the Cascades, the region under study in this volume. Ethnographic and historic accounts of native firing are common and began with the records from the Lewis and Clark expedition (Robbins 1999). Plateau peoples purposively used fires for the same ends as the Cascade and coastal peoples of the Northwest. The Kalispell, Nez Perce, Coeur d'Alene, and other Plateau peoples used fire to drive game (Boyd 1999, Lahren 1998, Robbins 1999), for improving browse for game (Marshall 1999, Palmer 1998), for improving berry crops (Marshall 1999), and for increasing camas and other root plants important to the Native American diet (Boyd 1999, Marshall 1999). Firing in lowland valleys and sage-grass steppe was used to promote the sprouting and growth of grasses at the expense of sagebrush. In the higher forested areas, fire was used to decrease lichen and mistletoe, increase grasses, and keep the ponderosa pine (*Pinus ponderosa*) stands free of firs and other invading tree species.

Fire History

It is likely that fire in the selected study area has been influenced by both natural and cultural factors. Currently, humans are a source of both accidental and intentionally set "prescribed" fires (Pyne 1982). Modern human behavior, as well as the ethnographic and historic evidence reviewed above, indicates that humans likely have a long history of promoting fires in this region. Written records indicate that climate is another major contributor to modern fires. In this area, lightening presently starts 21 to 40 fires per 400,000 hectares (million acres) per year (Martin et al. 1977). More recently, Kay (2007) reported that lightning fires approximately 60 miles (100 km) west of the study area in Okanogan and Wenatchee National Forests occur at a rate of 35 and 27 per 400,000 hectares per year,

respectively. Climate change can affect this source of ignition via an increase in storminess with a concomitant increase in lightening strikes, or via a decrease in effective moisture, since dry fuel is more flammable (Bergeron and Archambault 1993; Birks and Gordon 1985; Clarke 1988a; Pyne 1982, 1991; Swain 1973).

Long-term fire histories exist for selected parts of the Plateau and surrounding areas. At Smeads Bench Bog, in northwest Montana, Chatters and Leavell (1994) used lake sediments to provide microscopic pollen and charcoal. The authors demonstrated that fires in the surrounding hemlock forest occurred at an average 200-year interval over the past thousand years. They observed two different fire regimes, the first one spanning the thirteenth to sixteenth centuries A.D., showing a mean interval of 135 years. The second regime, spanning the sixteenth century A.D. to present, had a much longer return period of 283 years.

Microscopic charcoal studies from lake sediments in the ponderosa pine forests of Yellowstone Park demonstrate a more variable periodicity in fire. The Yellowstone study found that woodlands had prehistoric fire intervals that varied widely from 40 to 280 years at each of several local sites (Millspaugh and Whitlock 1995). In a regional overview of the Plateau (but using mostly data from the eastern Rockies), Chatters (1998) notes that reconstructed Holocene fire records tend to show that fires were most frequent and least severe in the period from 1,000 to 2,000 B.P. (Chatters 1998, Hemphill 1983, Mehringer et al. 1977b, Smith 1983). Chatters interprets this higher fire frequency as the product of purposive management to produce early-seral vegetation in order to enhance plant and animal foods important to native peoples (Chatters 1998).

Modern Climate

The modern climate in this region is primarily controlled by the configuration of two air masses—the Pacific subtropical high and Aleutian low. Changes in these air masses, mediated by the physiography of western North America, determine local weather conditions in the study area. The most frequent prevailing modern pattern is one in which the Aleutian low extends far south, increasing storminess and pushing moist maritime air masses to the Pacific Northwest on an eastward track. These cool, moist air masses do not deliver much precipitation to the northern Columbia Plateau because they first encounter the coastal ranges of the Olympic and Cascade Mountains as they make landfall (Ames 2000, Chatters 1998). Adiabatic cooling of air masses that try to ascend these coastal mountain ranges causes the moisture to fall as precipitation on the west side of the mountains, and results in a rain shadow on the east side of the mountains (Figure 2.4).

Even when the configuration of air masses changes, and the Aleutian low assumes a more northerly position, eastern Washington remains relatively arid. Under such conditions, the retraction of the Aleutian low allows arctic air masses to enter the Plateau region, bringing

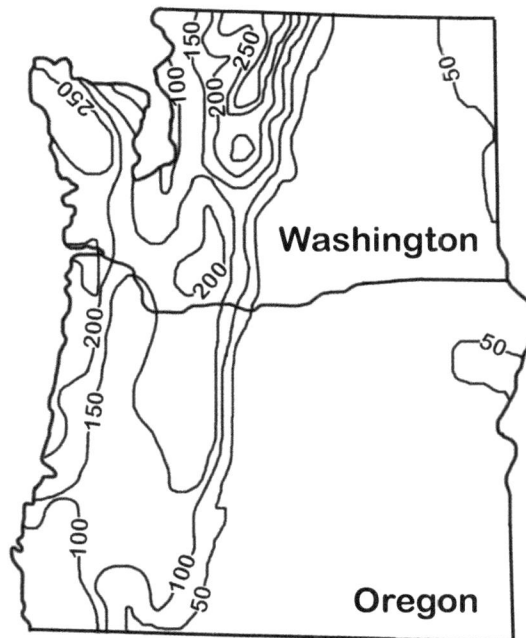

**Figure 2.4. Precipitation (cm/yr) in Washington and Oregon
(Based on Ames 2000, Figure 3)**

only dry continental air. Predictably, the study area and surrounding region are considered "semi-arid" with annual precipitation tending to average about 32 cm (Forbes 1987, Leeds et al. 1985).

Although precipitation is low on average, the variability of temperature and precipitation here varies seasonally and annually. The area typically experiences hot and relatively dry summers and cold, moister winters (Chatters 1998, Forbes 1987, Leeds et al. 1985). Instrument records illustrate this seasonal swing in temperatures, with mean January temperature averaging -4.0 °C, and July temperatures averaging 22.5 °C (U.S. Dept of Commerce, Weather Bureau in Forbes 1987).

Historic and instrumental records also attest to the degree of variability from year to year. Although annual precipitation averages 32±7 cm per year, the moisture delivered to this area can change quickly and unpredictably. One wet year of over 44 cm of recorded precipitation, for example, was followed immediately by a year of less than 18 cm of precipitation (Leeds et al. 1985).

Since the storm tracks determine the amount of precipitation entering the area, this variability in precipitation is determined by year-to-year changes in the extent and duration of the southern penetration of air masses such as the Aleutian low and the Pacific subtropical high. These weather systems, in turn, are controlled by the pattern and timing of ENSO (El Niño and La Niña) events (NOAA 2001, Taylor et al. 1998), and major fluctuations in the position of the jet stream. The result of this high interannual variability is that

conditions can change radically over a short period of time, and since the flora of semi-arid areas is inherently sensitive to moisture fluctuations, the growing conditions, productivity, composition, and presence of botanical communities will fluctuate drastically from year to year (Leeds et al. 1985).

Past Climate

Global climate reconstructions for the late Holocene (the period under consideration in this study) show several broad trends over this time period. The late Holocene, on the whole, is considered to be a relatively cool, moist period, preceded by a warm dry mid-Holocene trend (the Altithermal) which lasted from about 8,000 to 4,000 or 5,000 B.P. (Chatters 1998, Dalan 1985a, Mayewski et al. 2009, Sabin and Pisias 1996, Wright et al. 1993). The global trend beginning 4,000 or 5,000 B.P. is one of general cooling (the Neoglaciation), with several smaller-scale events imposed on the larger trend (Heusser et al. 1985, Sabin and Pisias 1996). Around 1,500 B.P. there was a small-scale global climate event resulting in a short-term relative warming (Heusser et al. 1980). Global precipitation trends are less commonly reconstructed, but precipitation seems to have remained more or less constant from 4,000 or 5,000 until 1,000 B.P., when a world-wide trend towards wetter conditions began (Heusser et al. 1980).

Imposed upon these general global late-Holocene trends described above are several smaller rapid climate change events (*sensu* Mayewski et al. 2004) that were low in both amplitude and frequency. One is a short-term excursion to warm conditions which supposedly took place between A.D. 900 and A.D. 1350, which is referred to as the "Medieval Warm Period" (MWP), the "Little Climatic Optimum," or the "Medieval Optimum" (Crowley and Lowery 2000, Grove and Switsur 1994, Hughes and Diaz 1994, Jones et al. 1999, Luckman 1994, Meese et al. 1994, Millspaugh and Whitlock 1995). More recently, the overall late Holocene global cooling trend culminated in a period of greatest relative cold. This low amplitude, low frequency cooling event is commonly referred to as the "Little Ice Age" or LIA, often cited as lasting from A.D. 1350-1860, although maximum cold temperatures may have been reached between 500 and 200 years ago (Bradley and Jones 1993, Graumlich and Brubaker 1986, Meese et al. 1994, Smith and Laroque 1996). It should be cautioned that the existence, nature, local expression, impact and utility of both the MWP and LIA are currently under debate in the literature (Broeker 2001, Hughes and Diaz 1994, Jones et al. 1999, Meese et al. 1994). A third and unnamed event has occurred in the past 100 to 200 years with yet another short-term change towards warmer conditions imposed on the general late Holocene cooling trend (Bradley and Jones 1993, Heusser et al. 1985). This recent warming trend is more widely acknowledged than the MWP and LIA as a truly global climatic trend.

Regional-scale climatic reconstructions presented for the Plateau over the late Holocene follow the generalized trends as described for the globe as a whole. Mountain glaciers in the Cascades on the western boundary of the Plateau retreated around 2,500 B.P., and re-advanced only around 900 to 1,000 B.P. (Ames 2000). Evidence from other sources such as cave spalling, erosion in alluvial fans, and freshwater mussel growth patterns agree with this evidence, pointing to a relatively warm and arid period in parts of the Plateau from around 2,500 to 1,000 B.P. (Chatters 1998).

Vegetation

Modern vegetation on the Southern Plateau is controlled by available moisture, and exhibits marked elevational zonation. Low elevations tend to be shrub steppe or mixed grass-shrub steppe with open forests restricted to higher elevations (Daubenmire and Daubenmire 1968, Daubenmire 1970).

Chatters (1998) and Nickman and Leopold (1985) have used palynological sources to produce a vegetation history for the Plateau region, redrafted and presented in Figure 2.5. At the end of the Pleistocene, vegetation characteristic of open woodland (Zone I in Figure 2.5) first established itself during the cool moist conditions directly following deglaciation. Around 10,000 B.P., a trend to warmer and drier conditions began. In response to this warming trend, vegetation converted first to a grassland (Zone II in Figure 2.5) and then to open shrubland (Zone III, Figure 2.5).

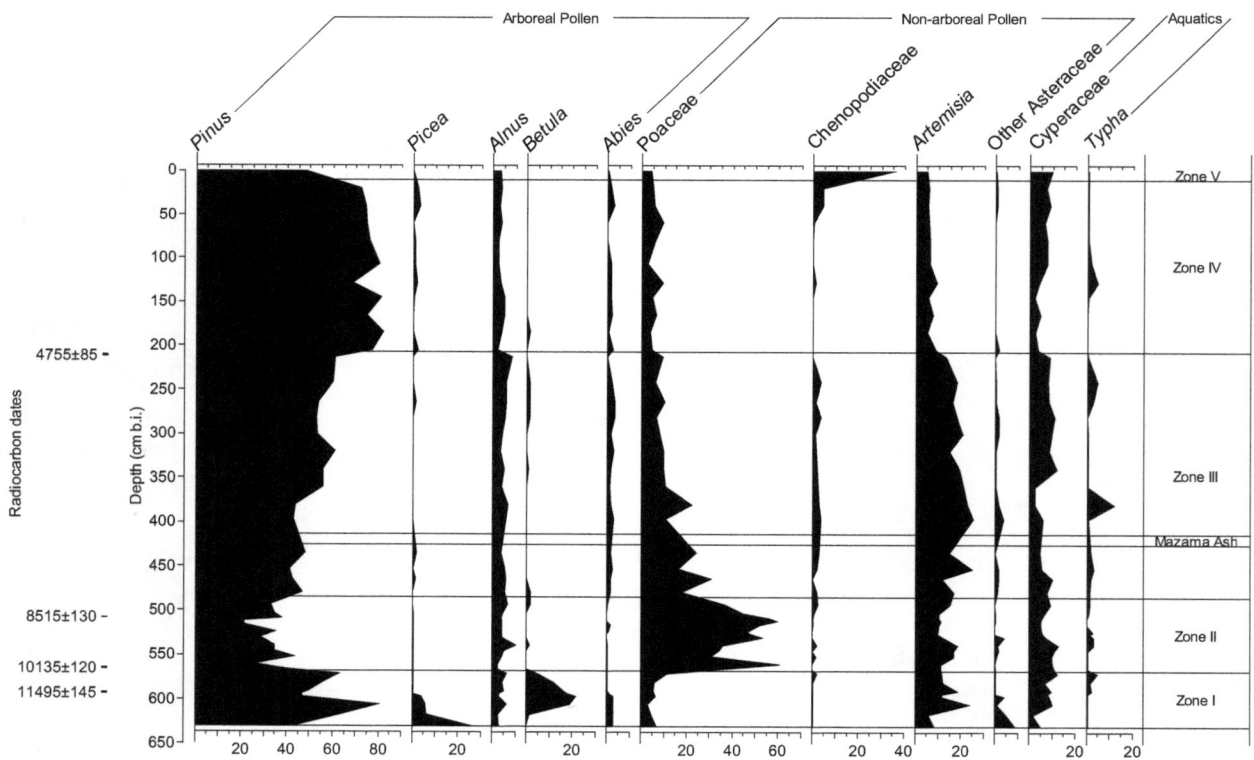

Figure 2.5. Pollen Diagram from Goose Lake
(Adapted from Dalan 1985a:117,118; dates from Nickmann and Leopold 1985)

Around the time of the Mazama ash, the driest conditions in the record prevail (top of Zone III), with vegetation characterized by sagebrush (*Artemesia*). By about 4,000 B.P., conditions become cooler and moister again (Mack et al. 1979, Nickmann and Leopold 1985). This mesic trend resulted in an increase in arboreal species (as seen in Zone IV in Figure 2.5). Palynologists with a millennial-scale perspective interpret this transition to a mixture of grass-shrub steppe and pine-dominated forest vegetation at about 4,000 B.P. as the establishment of the basic modern vegetation pattern in the area (Dalan 1985a, 1985b; Mack et al. 1979; Whitlock and Bartlein 1997), which has changed little since.

Others have examined the regional vegetation history at finer temporal scale and noted that the vegetation at 4,000 B.P. was only roughly equivalent to present patterns, since subalpine forest limits were actually lower than today. Forests were generally less open and more spatially extensive than today, indicating to many that the local climate may have been moister and cooler than today (Chatters 1998; Mack et al. 1977, 1978; Mehringer 1985).

Around 2,000 B.P. conditions and spatial patterns changed with subalpine forest margins moving upslope, and other forests thinning, contracting and being invaded by grasses (Chatters 1998; Mack et al. 1979, 1983). It could be said that by 4,000 B.P. the general modern vegetation patterns were established, with 2,000 B.P. marking the establishment of the exact zones and spatial patterns seen in historic times (Chatters 1998). Many researchers also note a very recent brief high-frequency trend in the regional vegetation in which Chenopodiaceae increases in density and ubiquity at the expense of pine and grass. This has been linked to the introduction of domestic herds and range grazing since the 1800s (Dalan 1985b, Davis et al. 1977). Additional information on vegetation history, along with the climate and fire histories, gives a regional context for understanding the archaeology of the study area. This information will be presented in the following chapters.

Chapter 3. Rufus Woods Lake Study Area (WA, USA)

The Southern Plateau provides an appropriate place to address the potential impacts of hunter-gatherer groups on the landscape. Additionally, the Southern Plateau has an existing demographic record that extends back seven thousand years at a 20- to 50-year resolution (Ames 2000, Chatters 1995b). In terms of scale, however, the Plateau is too vast to securely reflect human action (Delcourt and Delcourt 1988). Within the Plateau, however, a much smaller area immediately surrounding the Rufus Woods Lake provides a more suitable record. Besides having an available, pre-existing fine-scale demographic record, archaeological data for this area were collected under a single large-scale CRM project (Aikens 1988, Campbell 1985), producing "what is probably the region's best, large modern [archaeological] data set" (Ames 2000).

Prehistoric Human Population Record

Campbell (1989, 1990) produced a population proxy for a relatively small area within the Southern Plateau, based on archaeological phenomena that were compiled on the necessary mesoscale (extralocal scale). Campbell's study area covers a corridor along the Middle Columbia River of Washington State, from the Chief Joseph Dam to Grand Coulee Dam (Figure 2.3). Since the construction of the Chief Joseph Dam, the upstream portion of the river was converted to the reservoir referred to as Rufus Woods Lake or the Chief Joseph Reservoir (Campbell 1985, Ames et al. 1998). Rufus Woods Lake is located along the northern edge of the Columbia Basin, which is a relatively low-lying interior area of the Plateau, underlain by granitic batholiths and covered by a set of extensive basalt flows (Alt and Hyndman 1995). To the south, the land is relatively flat and covered with shrub-steppe; to the north, the area grades into the Okanogan

highlands. This general location is also referred to as the northern Columbia Plateau or the South-Central portion of the Plateau (Walker 1998).

Campbell (1989, 1990) produced a human demographic record which is decadal-to-centennial in scale. She used archaeological data from 77 site components (with 84 radiocarbon dates) in the study area to trace demographic changes that spanned the late prehistoric to early historic period along the corridor of the Columbia River from Chief Joseph Dam to Grand Coulee Dam. She used a series of several different proxy data sets to reconstruct prehistoric populations at 50-year intervals throughout the A.D. 950-1900 (1000-50 B.P.) time period (Campbell 1989, 1990).

Many different phenomena in the archaeological record are assumed to be related to past human population size. Archaeologists commonly use the number of sites or the area (m^2) of habitation as proxies for past population size (Campbell 1989, 1990; Ramenofsky 1987). In her study, Campbell (1989, 1990) used traditional measures of the number and area of occupations, but also included measures of the production of food waste and features (such as hearths) in her analysis. Examples of her proxy measure plots are given in Figures 3.1 and 3.2, which show the number of features and amount of animal-bone (food waste) produced per unit time. These proxies, like those used by other archaeologists, basically measure population size through quantification of human impact on the archaeological record itself; the higher the rate of production of archaeological phenomena, other things being equal, the greater the assumed associated population size.

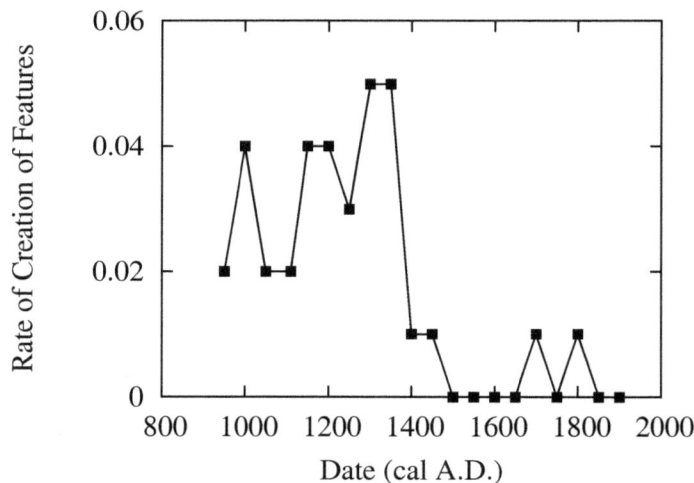

**Figure 3.1. The Rate of Occurrence of Features per 50-year Interval
(Adapted From Campbell 1989:172)**

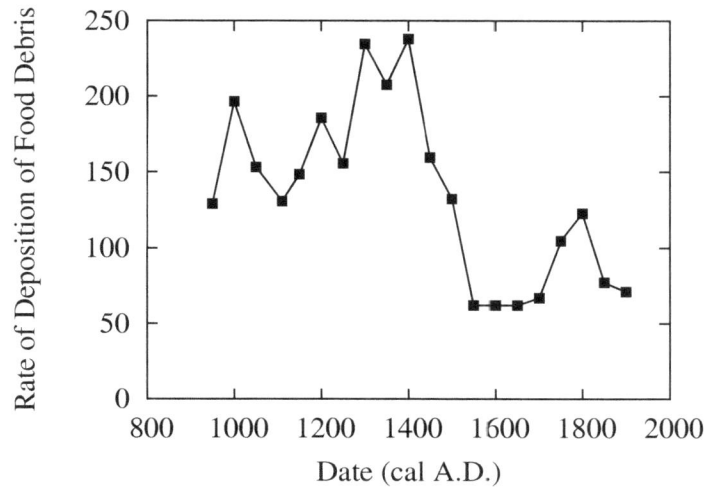

**Figure 3.2. The Rate of Deposition of Bone Debris (Food Waste) per 50-year Interval
(Adapted From Campbell 1989:174)**

Table 3.1 lists Campbell's (1989) data for six variables that she considered important: the number of components, the habitation area, the average #/m²/yr of bone debris, the g/m²/yr of bone debris, the g/m²/yr of shell debris, and the number of features created during each of her 50-year time intervals. These data were gleaned from site records for all recorded sites inhabited from A.D. 950 to 1900. These sites were documented primarily during reclamation projects relating to the construction of dams or draw-downs of reservoirs in the area. This brings strengths and weaknesses. Because most of the sites were excavated by a single team using a constant set of methods, data from many different sites are comparable. However, there is also a bias towards archaeological sites near the river, which tend to be winter village sites. Although the more ephemeral upland sites (such as hunting camps, quarries, berry and root gathering camps) were not given as much attention during the reservoir projects, a large number of them are known and documented from the Rufus Woods Lake area. In fact, the areas away from the river in this area are better known archaeologically than the uplands in the rest of the Southern Plateau (Ames et al. 1989).

Temporal control for Campbell's population surrogates was derived mainly from radiocarbon dating, with support from artifact seriations and the occasional presence of historic trade goods of known age. Campbell calibrated all radiocarbon dates, and found that this established a chronology that, even with the inherent uncertainty of radiocarbon dating, produced a population proxy with a 50-year resolution. Figure 3.1 shows the occurrence of pit, hearth and oven features, while Figure 3.2 plots the production of bone (food) debris, as calculated by Campbell (1989). These graphs show relatively high population values early in the record, a drop around A.D. 1400, a slight increase in the 1800s and a final drop by A.D. 1900. Confidence in Campbell's

surrogate variables comes from her demonstration that the proxy indicators she used are in agreement with each other, and thus are measuring a single underlying variable—presumably population density. Table 3.2 lists the correlations between the six population surrogates used by Campbell (1989). These proxies, in general, were highly correlated with each other and with the composite population proxy constructed for this study. All measures except the production of shell debris were significantly (at the 0.05 level) and highly positively (r ranging from +0.489 to +0.909) correlated with the composite population index. Obviously, these variables are all measuring the same basic underlying phenomenon.

Additional support for Campbell's surrogates can be found in their agreement with other independent measures of population. Campbell's proxies compare favorably with the prehistoric population reconstructions given for the Plateau as a whole (Figure 2.2). The demographic reconstruction by Chatters (Chatters 1995b, Ames 2000) exhibits the same timing and variance as Campbell's data. Looking at just the past thousand years of Chatter's proxy, values start off high a thousand years ago and decrease rapidly between 300 and 500 B.P. Populations partially recover, and drop again some 100 years before present. This is the same pattern seen in Campbell's plots. Ames' reconstruction (Figure 2.2) also documents the precipitous decline around 500 B.P., but his plot cannot be used to evaluate more recent trends as it ends by 300 or 400 B.P. (Ames 2000).

Further confirmation of the reliability of Campbell's paleodemographic work comes from its similarity to trends known from historic and ethnohistoric sources. Boyd (1985, 1990; Campbell 1989) documented two of the same events as seen in the archaeological population proxy from the Rufus Woods Lake area. Boyd also observed a partial population recovery in the early

Table 3.1. Paleodemographic Record—Population Surrogates Used

Temporal Interval cal A.D.	Interval Midpoint cal A.D.	# of Components	Total Area m^2	Average Bone #/m^2/yr	Average Bone g/m^2/yr	Average Shell g/m^2/yr	Average Features #/m^2/yr
1875-1925	1900	8	11202.00	71.19	13.49	0.06	0.00
1825-1875	1850	9	30402.00	77.26	14.56	0.70	0.00
1775-1825	1800	13	37902.00	122.71	39.31	50.66	0.01
1725-1775	1750	17	47239.00	104.57	38.46	32.22	0.00
1675-1725	1700	14	42729.00	66.93	32.55	24.31	0.01
1625-1675	1650	17	57903.00	62.07	21.46	18.32	0.00
1575-1625	1600	17	57903.00	62.07	21.46	18.32	0.00
1525-1575	1550	17	57903.00	62.07	21.46	18.32	0.00
1475-1525	1500	18	61975.00	132.27	47.31	3.13	0.00
1425-1475	1450	26	76111.00	159.49	49.91	14.93	0.01
1375-1425	1400	20	66404.00	237.79	53.47	18.63	0.01
1325-1375	1350	22	49790.00	207.44	54.06	20.25	0.05
1275-1325	1300	27	79655.00	234.71	57.87	21.31	0.05
1225-1275	1250	23	74609.00	155.53	37.92	21.31	0.03
1175-1225	1200	22	75014.00	185.63	54.12	21.31	0.04
1125-1175	1150	18	56662.00	148.35	49.31	14.93	0.04
1075-1125	1110	18	53020.00	130.60	86.60	14.93	0.02
1025-1075	1050	21	55732.00	153.05	87.12	11.80	0.02
975-1025	1000	23	64324.00	196.43	77.76	10.73	0.04
925-975	950	31	101437.00	129.08	43.72	13.03	0.02
		$\bar{x} = 19.05$	$\bar{x} = 57895.80$	$\bar{x} = 134.96$	$\bar{x} = 45.10$	$\bar{x} = 17.46$	$\bar{x} = 0.02$
		$s_x = 5.68$	$s_x = 19438.24$	$s_x = 57.47$	$s_x = 21.63$	$s_x = 11.07$	$s_x = 0.02$

Data are taken from Campbell 1989:166
Interval midpoints, averages, and standard deviations added later

Table 3.2. Correlation Matrix for Campbell's Population Surrogates

	# Components	Area (m^2)	Bone (#/m^2/yr)	Bone (g/m^2/yr)	Shell (g/m^2/yr)	Features (#/m^2/yr)	Composite Population Index	
# Components	1.00 -	.932 .000	.632 .003	.525 .017	.057 .810	.577 .008	.632 .003	r sig*
Area (m^2)		1.00 -	.489 .029	.369 .109	.058 .808	.409 .073	.489 .029	r sig*
Bone (#/m^2/yr)			1.00 -	.674 .001	.084 .726	.756 .000	.909 .000	r sig*
Bone (g/m^2/yr)				1.00 -	.023 .923	.580 .007	.673 .001	r sig*
Shell (g/m^2/yr)					1.00 -	.096 .686	.084 .724	r sig*
Features (#/m^2/yr)						1.00 -	.756 .000	r sig*
Composite Population Index							1.00 -	r sig*

*r is Pearson's r, significance is two-tailed

16

historic era, followed by a recent decline around A.D. 1850-1900. Because of the agreement between various measures of past population, Campbell's surrogates can be considered quantitative measures of prehistoric population, providing a solid base-line for interpreting the climate, vegetation, and fire-frequency records to be generated in the current study.

For the current study, information from all of Campbell's surrogates was combined to form a single composite index of prehistoric population. Each of the six proxies was normalized into a Z-score in order to remove bias and make the measurements comparable to each other. The variables were averaged together for each 50-year period once each observation was normalized. To standardize a particular variable, the mean value and standard deviation were first computed (Table 3.1). For example, the average number of components for all 50-year periods was 19.05, with a standard deviation of 5.68. For each interval, the average was subtracted from the actual observed number of components, and the result was divided by the standard deviation, as shown by the computations in Table 3.3. For the A.D. 1875-1925 interval, for example, 8 components were observed. This was converted to a normal score by computing, for this example, (8-19.05)/5.68, producing a normalized value

of -1.95. Later, all normalized scores from A.D. 1875-1925, one from each proxy variable, were averaged together and this was used as composite index (Table 3.4). This was done for each 50-year interval, in turn (Table 3.4).

The resulting composite population proxy is plotted in Figure 3.3. The composite population index shows a high density of people early in the record (A.D. 950-1400), a precipitous drop sometime after A.D. 1400, a partial population recovery, and then a final decline during historic times. This composite index shows the same trends as the individual proxies (Figures 3.1 and 3.2), and also upholds Campbell's conclusions that a drastic population crash occurred around the time of Contact.

Campbell (1989, 1990) went further to conclude that the population decline that she documented was due to external factors, specifically introduced diseases. Statistical analysis of the composite index demonstrates that human population values are not historically constrained: they are not being controlled by intrinsic factors. This can be shown by correlating population size, at any given time, to the size of the population 50, 100, 150, or 200 years in the past.

Table 3.3. Standardization of a Variable (# of Components)

Temporal Interval cal A.D.	Interval Midpoint cal A.D.	# of Components	Computation of Normalized Z-score (#components-x̄)/s_x	Resulting Normalized Z-score
1875-1925	1900	8	(8-19.05)/5.68	-1.95
1825-1875	1850	9	(9-19.05)/5.68	-1.77
1775-1825	1800	13	(13-19.05)/5.68	-1.07
1725-1775	1750	17	(17-19.05)/5.68	-0.36
1675-1725	1700	14	(14-19.05)/5.68	-0.89
1625-1675	1650	17	(17-19.05)/5.68	-0.36
1575-1625	1600	17	(17-19.05)/5.68	-0.36
1525-1575	1550	17	(17-19.05)/5.68	-0.36
1475-1525	1500	18	(18-19.05)/5.68	-0.18
1425-1475	1450	26	(26-19.05)/5.68	1.22
1375-1425	1400	20	(20-19.05)/5.68	0.17
1325-1375	1350	22	(22-19.05)/5.68	0.52
1275-1325	1300	27	(27-19.05)/5.68	1.40
1225-1275	1250	23	(23-19.05)/5.68	0.70
1175-1225	1200	22	(22-19.05)/5.68	0.52
1125-1175	1150	18	(18-19.05)/5.68	-0.18
1075-1125	1110	18	(18-19.05)/5.68	-0.18
1025-1075	1050	21	(21-19.05)/5.68	0.34
975-1025	1000	23	(23-19.05)/5.68	0.70
925-975	950	31	(31-19.05)/5.68	2.10

$$\bar{x} = 19.05$$
$$s_x = 5.68$$

Data are taken from Campbell 1989:166
Interval midpoints, averages, standard deviations, and Z-scores added

Table 3.4. Z-scores and Composite Population Index

Temporal Interval cal A.D.	Interval Midpoint cal A.D.	Column A Z-score # Components	Column B Z-score for Area	Column C Z-score for Bone ($\#/m^2/yr$)	Column D Z-score for Bone ($g/m^2/yr$)	Column E Z-score for Shell	Column F Z-score for Features	Column G Composite Population Index*
1875-1925	1900	-1.95	-2.40	-1.11	-1.46	-1.57	-0.97	-1.58
1825-1875	1850	-1.77	-1.41	-1.00	-1.41	-1.51	-0.97	-1.35
1775-1825	1800	-1.07	-1.03	-0.21	-0.27	3.00	-0.42	0.00
1725-1775	1750	-0.36	-0.55	-0.53	-0.31	1.33	-0.97	-0.23
1675-1725	1700	-0.89	-0.78	-1.18	-0.58	0.62	-0.42	-0.54
1625-1675	1650	-0.36	0.00	-1.27	-1.09	0.08	-0.97	-0.60
1575-1625	1600	-0.36	0.00	-1.27	-1.09	0.08	-0.97	-0.60
1525-1575	1550	-0.36	0.00	-1.27	-1.09	0.08	-0.97	-0.60
1475-1525	1500	-0.18	0.21	-0.05	0.10	-1.29	-0.97	-0.36
1425-1475	1450	1.22	0.94	0.43	0.22	-0.23	-0.42	0.36
1375-1425	1400	0.17	0.44	1.79	0.39	0.11	-0.42	0.41
1325-1375	1350	0.52	-0.42	1.26	0.41	0.25	1.80	0.64
1275-1325	1300	1.40	1.12	1.74	0.59	0.35	1.80	1.17
1225-1275	1250	0.70	0.86	0.36	-0.33	0.35	0.69	0.44
1175-1225	1200	0.52	0.88	0.88	0.42	0.35	1.25	0.72
1125-1175	1150	-0.18	-0.06	0.23	0.19	-0.23	1.25	0.20
1075-1125	1100	-0.18	-0.25	-0.08	1.92	-0.23	0.14	0.22
1025-1075	1050	0.34	-0.11	0.31	1.94	-0.51	0.14	0.35
975-1025	1000	0.70	0.33	1.07	1.51	-0.61	1.25	0.71
925-975	950	2.10	2.24	-0.10	-0.06	-0.40	0.14	0.65

*average of all Z-scores for six potential population surrogates.
For each row, column G is computed as G = (A+B+C+D+E+F)/6

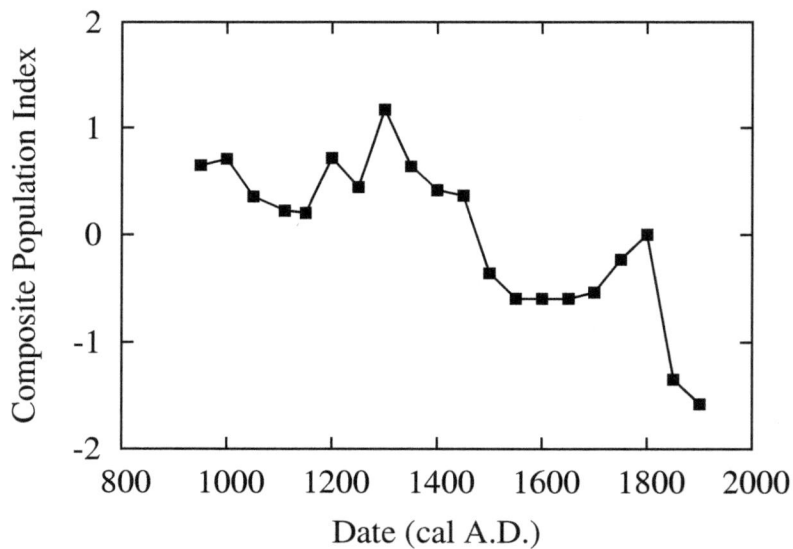

**Figure 3.3. Composite Population Index from Study Area
(Calculated from data in Campbell 1989)**

Table 3.5 lists the correlations between population size and past populations, with lags of one to ten intervals (50 to 500 years) introduced into the correlation model. When examined for autocorrelations (Table 3.5), the population index shows a strong dependence on values of itself one time interval (50 years) in the past. Prehistoric population in the Rufus Woods Area, however, is not highly correlated with itself over longer time frames. This indicates that population sizes are not highly constrained by demographic events in the distant past, and some external factor is influencing the system. Whether this external factor is disease, as hypothesized by Campbell (1989, 1990), cannot be addressed by the current analysis. Currently, there is an ongoing debate over whether introduced diseases had a simultaneous and similarly disruptive effect on all New World populations (Denevan 1996, Kealhofer and Baker 1996). In many areas, local trajectories have been shown to be unique as different settings and cultures responded in unique ways to alien pathogens (Kealhofer and Baker 1996). Some have even argued that long-term depopulation of Native American communities was due, in large part, to small changes in the demographic regime rather than from epidemic diseases themselves, arguing that the direct impacts of European colonization (warfare, ecological changes, social and settlement pattern changes) were the primary reason for population declines (Thornton 1997).

Paleoeconomic Information

Although the size of the population in this area over the past thousand years is a critical variable, the resource procurement activities of those people are important as well. Some information can be gained on lifeways and economies from ethnographic and historic records. Written records for this area are of limited utility, however, since the historical records do not extend far back in time; the first face-to-face contact between Native American groups and Euro-Americans in this area occurred during the Lewis and Clark expedition of 1805. The number of Euro-Americans (and written records) in this area remained low until the late 1850s, and it was not until after 1900 that the first ethnographic work began (Campbell 1989, Leeds et al. 1985, Lohse and Sprague 1998).

The seminal ethnographic account of Plateau peoples is that done by Ray in 1932, supplemented later by work published by Spier in 1939. These ethnographies indicate that the study area was occupied by Interior or "Middle-Columbia-River" Salishan speakers (Miller 1998) practicing a central-based "intensive" subsistence system, relying on the collection of plant foods and game with a focus on anadromous fish, primarily salmon (Hunn 2000, Leeds et al. 1985). These people spent the winter months in permanent semi-subterranean dwellings in villages located in the floodplain of the river. During summer months, groups would fission and move around the landscape after plant and game resources, using more mobile mat lodges as shelter (Kirk and Daugherty 2007).

Prehistoric data from this area is mainly derived from cultural resource management projects that have been

Table 3.5. Composite Population Index Autocorrelations
(Correlations between Population Index and Future Values of Itself)

Significant Autocorrelations

	Correlation (Pearson's *r*)	Significance (*p*)
1 lag (50 years)	0.782	< 0.0001

Insignificant Autocorrelations

	Correlation (Pearson's *r*)	Significance (*p*)
2 lags (100 years)	0.564	< 0.015
3 lags (150 years)	0.505	< 0.039
4 lags (200 years)	0.428	< 0.098
5 lags (250 years)	0.396	< 0.144
6 lags (300 years)	0.434	< 0.121
7 lags (350 years)	0.518	< 0.070
8 lags (400 years)	0.473	< 0.121
9 lags (450 years)	0.461	< 0.153
10 lags (500 years)	0.365	< 0.300

carried out as part of the hydroelectric and irrigation projects along this part of the Columbia River Basin, mainly between the 1940s and 1980s (Ames 2000, Campbell 1989). Data from these projects, paired with other sources, have resulted in the following narrative, which is a consensus overview of this area's human land use. Throughout the prehistoric era, the area that now lies along the Columbia River between the Chief Joseph and Grand Coulee Dams (Figure 2.3) was populated by people practicing a hunting and gathering way of life. In the early and middle Holocene (approximately 8,000 to 2,000 B.P.) the people inhabiting this area practiced an "extensive," "wandering" form of hunter-gatherer life (Leeds et al. 1985). That is, in the warm dry Altithermal period directly preceding the focus of this study, the people in the study area were involved in a generalized strategy in which a wide variety of low-, medium-, and high-ranked resources such as small game, roots, shellfish, and seeds were collected as groups of foragers moved often across the landscape. Culture-historical units that correspond to or fall into this period include the Kartar, Hudnut, Vantage, and Quilomene Bar Phases (although these have different phase names in other local areas of the Plateau, as well as having been given new "synthetic" names in recently published overviews of Plateau prehistory) (Ames 2000, Ames et al. 1998, Chatters 1998, Galm 1994, Leeds et al. 1985). The consensus opinion among archaeologists is that this span of time represents a period of very low population density in the Columbia Basin as a whole (Ames 2000, Bense 1972, Chatters 1995b, Leeds et al. 1985). During this period, 8,000 to 2,000 B.P., groups were very mobile. People did not stay long in any one place, nor did they return year after year to the same locations. There were few capital investments made to archaeological sites during this time. Storage pits, hopper mortars, substantial habitation structures, and large aggregations of families did not occur in this area until later (Kirk and Daugherty 2007). Even so, there is evidence for extensive marine shell and obsidian trade networks, linking Plateau groups with other regions (such as Puget Sound, the Great Basin, and possibly Northern California) dating to this period, and possibly earlier (Galm 1994).

The focus of human use of this area's resources shifted around 2,000 B.P. (Leeds et al 1985, Galm 1994). Subsistence systems became centrally-based as people occupied villages near the Columbia River for part of the year, and spent the rest of the time moving across the landscape. Again, as during the previous six thousand years, people used a wide variety of low-, medium-, and high-ranked resources. Although taking the same range of resources, the people of this period were more numerous and hence put more pressure on the resources being taken. Also, congregations of populations in one area (winter villages) resulted in a more intensive exploitation of firewood in one place over one particular season. These patterns are associated with the Early and Middle Cayuse Phases. This period of time is assumed to have higher human population density than the period before it (Leeds et al. 1985). Semi-subterranean pit house structures were constructed and maintained, to be

revisited year after year. Storage pits, hopper mortars, and other permanent features became common as foods were harvested in bulk and preserved for later use (Chatters 1995, Kirk and Daugherty 2007). The later part of this time period witnessed an average increase in the size of structures, many of them presumed to house multiple families (Ames et al. 1998). It is also during this period that the bow and arrow were introduced into the region (Kirk and Daugherty 2007), and trade with groups outside the Plateau increased (Galm 1994).

From 350 to 150 years ago, corresponding in time to what is referred to as the Late Cayuse Phase, subsistence and land use differed from all of these earlier patterns. Starting about 350 years ago, native peoples began to restrict the range of resources they were taking. Instead of focusing on a wide range of low-, medium-, and high-ranked resources, people of this time period increased their reliance on a single high-ranked resource—salmon. Hunn (2000) cautions that this was a relative increase, and that in terms of caloric contributions, roots were still providing over half the winter food intake for Plateau peoples. A few intrusive behaviors began during this period, with new resources and practices introduced through direct and indirect contact with non-native peoples. Although face-to-face interaction with Europeans and Euro-Americans was relatively late in eastern Washington, indirect effects spread through interaction between native groups along established trade routes (some of which had been operating for at least 8,000 years [Galm 1994]). By at least A.D. 1730, domestic horses were obtained by Plateau people from other Native American groups (probably Plains groups), beginning the practice of keeping grazing animals (Campbell 1989, Davis et al. 1977, Hunn 2000), a trend that blossomed in a later period when Euro-American cattle ranchers came in numbers to the Plateau (Weddell 2001). Also traveling along established Native American trade routes, introduced diseases preceded face-to-face contact between Native Americans and Euro-Americans on the Plateau, with local smallpox epidemics dating back before 1780 (Boyd 1985, Campbell 1989, Hunn 2000). The early 1800s (approximately A.D. 1811 to 1850) brought a brief period of trapping for the fur trade, and an influx of those seeking to do missionary work in the area (Hunn 2000).

Land use changed again around 150 years ago with the introduction of Euro-American land use practices and Euro-American populations (Davis et al. 1977, Mack 1988, Weddell 2001). By the 1850s and 1860s the area was inundated with both Euro-American settlers and with many of their land use behaviors such as mining, logging, agriculture, and herding. These activities changed both the cultural and natural landscape, as great changes occurred in fire frequency, the composition of the biota, and the rate of erosion (Davis et al. 1977, Mack 1988). The introduction of large numbers of sheep and cattle resulted in overgrazing, which promoted seral plant communities and invading alien species (Davis et al. 1977, Mack 1988, Weddell 2001). By 1910, wheat fields and orchards had destroyed most local vegetation, eliminating the steppe vegetation that had played a large role in Native American lifeways (Mack 1988, Weddell 2001). Reports from botanical surveys that predate 1910 have been used as a baseline for describing the taxa, botanical associations, and spatial patterning of vegetation at the time of Contact in this area (Daubenmire 1970, Mack 1988, Weddell 2001).

Study-Area Vegetation

Modern Vegetation

As with the greater Plateau region, vegetation in the study area is controlled by available moisture, which is determined by the climatic balance between precipitation and evaporation. Within the study area, available moisture and the resulting plant communities are largely mediated by elevation (Chatters 1998, Leeds et al. 1985). Large areas in the lower elevations are covered by shrub steppe composed primarily of big sagebrush (*Artemesia tridentata*) and cheat grass (*Bromus tectorum*), an introduced species (Leeds et al. 1985). This is an extremely recent plant association, and replaced the similar established historically recorded shrub steppe communities composed of sagebrush (*Artemisia* spp.) and a variety of native bunch-grass species (mostly *Festuca idahoensis* or *Agropyron spicatum*, with some *Stipa comata*, *Stipa thurberiana*, *Poa cusickii* or *Sitanion hystrix*) (Leeds et al. 1985, Mack 1988).

Vegetation zones are shown in Figure 3.4. Areas of shrub-steppe cover the slopes leading down to the Columbia River from the north and south (river right and river left), as well as the flat-lying areas directly above the slopes of the river valley (Zones I, IIR, IIL and IIIL in Figure 3.4) (Dalan 1985a, Daubenmire 1970, Daubenmire and Daubenmire 1968, Chatters 1998, Franklin and Dyrness 1973, Leeds et al. 1985). Although this grass-sagebrush community accounts for most of the acreage in the areas close to the Columbia River, smaller areas containing slightly different plant associations exist within this broader vegetation zone, allowing for plants like bitterbrush (*Purshia*) and rabbit brush (*Chrysothamnus)* to become locally dominant in patches. Small freshwater lakes dispersed throughout the grass-shrub steppe, for example, support a small community of fringing aquatic vegetation (such as cattail, rushes, and sedges—*Typha*, *Juncus*, and *Carex*) and some of them support a few arboreal individuals (such as *Salix* and *Betula*). The shrub steppe zone is also transected by streams running through narrow canyons that support a mixture of mesic trees and shrubs such as alder (*Alnus*), birch (*Betula*), and poplar (*Populus*). The exact taxonomic composition of the shrub steppe is also modified the nature of the substrate, with the steeply sloping basaltic lithosols directly south of the river (in Zone IIL) supporting relatively more economically important root crops such as balsamroot (*Balsamorhiza sagittata*), bitter-root (*Lewisia rediviva*), and lomatium (*Lomatium*) than the slope on the north side of the Columbia River.

North of the Columbia River, yet still within the study area, lies a higher elevation area known as the Okanogan highlands that supports a more enclosed type of vegetation—an open coniferous forest, dominated by ponderosa pine (*Pinus ponderosa*) with an understory similar to the shrub steppe (Mack 1988). As expected, modern pollen spectra from this open woodland is dominated by pine and grass pollen (Mack et al. 1978). The higher areas are dotted with a scatter of small patches of hardwoods/deciduous trees (Zone IIIR in Figure 3.4), changing to a mixed Douglas fir (*Pseudotsuga menziesii*) and grand fir (*Abies grandis*) forest at the highest elevations along the boundaries of the study area (Zone IV in Figure 3.4).

Past Vegetation

There are two published pollen diagrams from this area, from Goose and Rex Grange Lakes, in Okanogan and Douglas Counties. Figure 3.5 is an area map showing the location of these lakes and Appendix B has further location information. The pollen diagram from Goose Lake (Figure 2.5) covers the entire Holocene, and has radiocarbon and tephrochronologic temporal controls (Nickmann and Leopold 1985). The second pollen diagram from Rex Grange Lake, given in Figure 3.6, focuses on recent times, and is undated (Dalan 1985b).

These two studies show that the vegetation change during the Holocene in the study area followed the same major path of vegetation change in the intermontane region as a whole (presented in Chapter 2). The transition from the Pleistocene to the Holocene (from Zones III to IV in Figure 2.5) is marked by a sharp decrease in pine and birch, with a sharp increase in grass, probably indicating the onset of warmer, drier conditions. From 7,800 to 6,700 B.P., (Zone V in Figure 2.5) grasses decrease while sagebrush and pines increase, which has been interpreted as possible evidence of increased available moisture in the area (Nickmann and Leopold 1985). The post-Mazama period is treated as a single zone. Within the post-Mazama period, sagebrush increases dramatically directly after the deposition of the Mazama ash. Later, grasses and chenopods increase at the expense of the sagebrush, and from 4,000 B.P. to the present day, ponderosa pine is dominant (Mack et al. 1976). Dalan (1985a) examines the upper zone in more detail (Figure 3.6), and gives a more finely-resolved record of changes in grass, sagebrush, and chenopods in the upper 19 cm of the Rex Grange core.

Although a detailed record of the late Holocene at Rex Grange Lake exists, it cannot be used because it has no associated radiocarbon dates or other chronological control (Dalan 1985a). The longer pollen record from Goose Lake (Dalan 1985a, Nickmann and Leopold 1985) is also inadequate for the purposes of this study because it allows vegetation to be viewed on a millennial, or at best centennial time-scale. In the existing published sequence, the post-Mazama record consists of a series of 18 samples. The average amount of time represented by the interval between samples is 372 years. Considering the resolution, the Goose Lake and Rex Grange Lake pollen records cannot be meaningfully compared to the

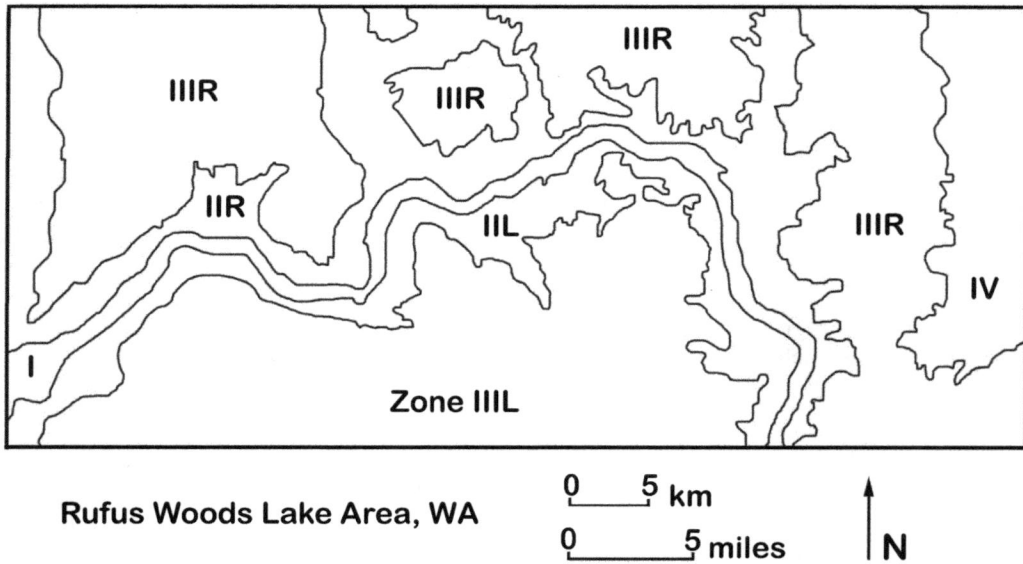

Rufus Woods Lake Area, WA

0 ____ 5 km

0 _____ 5 miles

N

Figure 3.4. Vegetation Zones in the Rufus Woods Lake Study Area
(Modified from Dalan 1985a and Leeds et al. 1985)

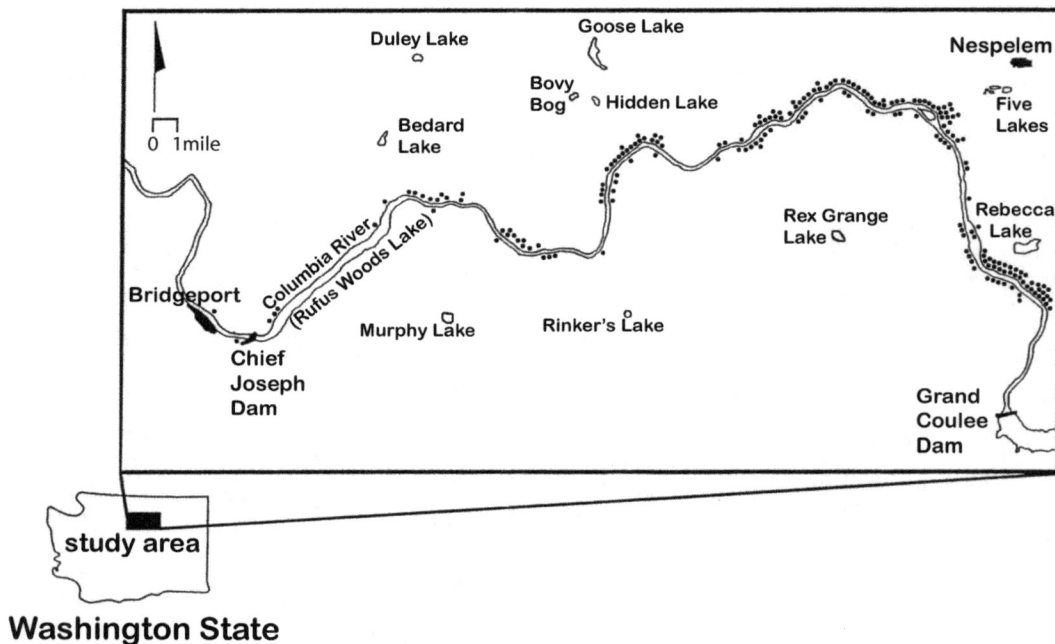

Washington State

Figure 3.5. Location of Study Area
Showing the Basins Discussed in the Text

Modern towns and modern dams are shown for reference.
Dots indicate locations of archaeologically-known prehistoric habitation (winter village) sites.
(Adapted from Campbell 1985, with information from Dalan 1985b, Forbes 1987, Nickmann and Leopold 1985, USGS 7.5' Belvedere Quad 1990, USGS 7.5' Boot Mountain Quad 1990, USGS 7.5' Bridgeport Point Quad 1990, USGS 7.5' Joe Lake Quad 1990, USGS 7.5' Sanderson Creek Quad 1990, USGS 7.5' Stubblefield Point Quad 1990, and USGS 7.5' Trefrey Canyon Quad 1990).

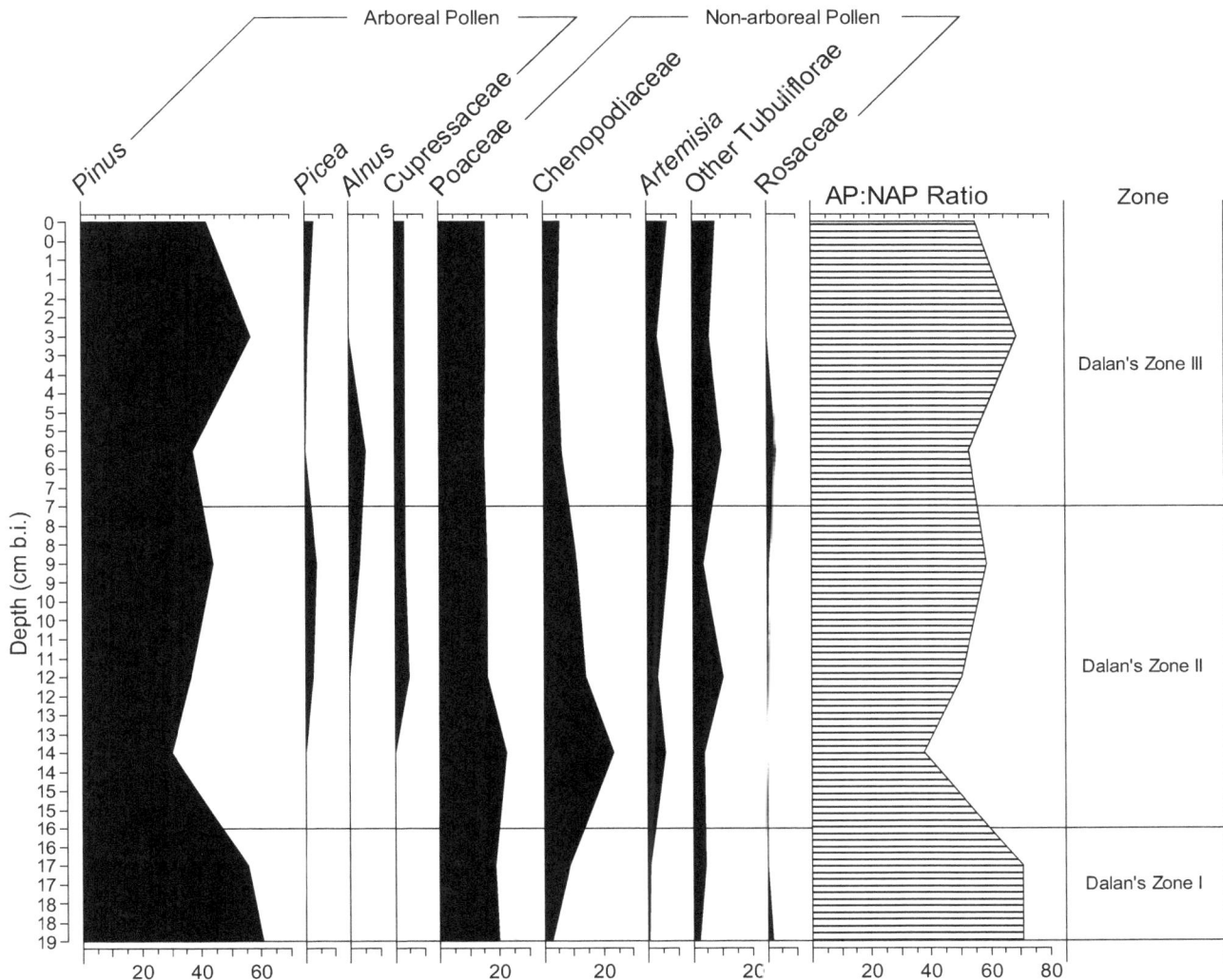

**Figure 3.6. Pollen Diagram for Rex Grange Lake
(Adapted from Dalan 1985b:524,525)**

paleodemographic or other records used in the current research (see Dalan 1993, Linse 1993, and Stein 1993 for similar scale problems). To remedy this situation, a new pollen record must be created for comparison to Campbell's data, matched to the desired scale and period of interest. This palynological record is presented and discussed in Chapter 6.

Study-Area Climate History

Although the review of pollen data, above, indicates some major climatic changes in the study area, these cannot be used as a proxy for climate for two reasons. First, as detailed above, the existing records are insufficient in temporal resolution. Second, the use of pollen-derived climate estimates for this study would be inherently circular. If pollen is used as a proxy for vegetation, and the resulting vegetation history is used in turn as a proxy climate record, then it is necessarily the case that the vegetation and climate records created in this manner will be based on the same underlying dataset.

It would not be mathematically or logically valid to use a pollen-derived climate estimation to explain a change in vegetation that is derived from the same pollen dataset; a truly independent line of climatic evidence is required.

Previous work in the study area includes a master's thesis aimed at creating a climatic reconstruction for the Rufus Woods Lake area. In this work, Forbes (1987) presented an oxygen isotope record from carbonates of Goose and Duley Lakes (Figures 3.7 and 3.8, respectively). Forbes determined that isotope records from these lakes were tracking increases and decreases in available moisture, and were sensitive to climate changes on a mesoscale. Unfortunately, Forbes' records span the entire post-Pleistocene era and are sampled at approximately 140-200 year intervals making his results inappropriate for comparison to the established paleodemographic record as well. Because of this, a new set of lacustrine samples were taken and oxygen isotope assays were run on to provide a new climate proxy for the study area (see Chapter 5).

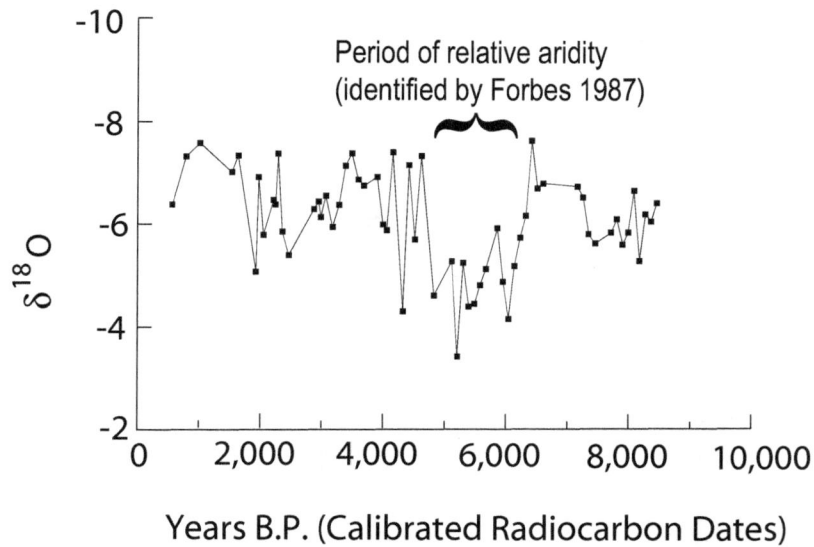

**Figure 3.7. Oxygen Isotope Diagram for Goose Lake
(Based on Data in Forbes 1987)**

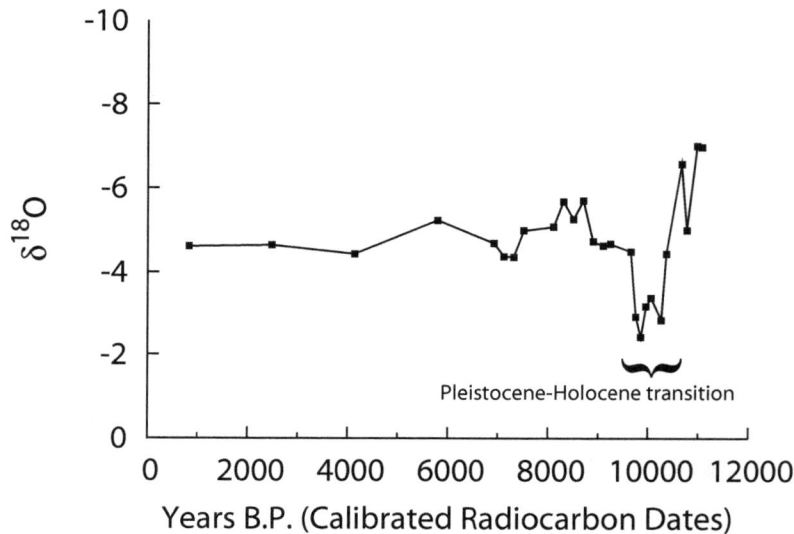

**Figure 3.8. Oxygen Isotope Diagram for Duley Lake
(Based on Data in Forbes 1987)**

Study-Area Fire History

There are no existing prehistoric fire history records that span the past millennium in the study area itself, although there are some sources of shorter-term historic information. Records indicate that from approximately A.D. 1890 to 1970, fires were suppressed (Dell 1977) on this part of the Columbia Plateau. This caused deadwood and similar fuels to build up on the landscape, allowing for large, intense, and destructive fires to potentially occur. Before fire suppression, it seems that fires in the arid inland Northwest were relatively common. The

importance of fire in this area was re-discovered in the late 1960s and early 1970s, and the study area, in particular, was a focus for research and a new plan for using prescribed burns to better manage the area (Martin et al. 1977). Research affirmed that fire is an important factor determining the composition and areal extent of vegetation types, especially in semi-arid environments such as this. Although climate and elevation are major determinants of the spatial boundaries and composition of vegetation zones found near Rufus Woods Lake, the size and frequency of fires also control the extent and openness of wooded versus grassy acreage. Studies of fire scars in standing forests and stumps from outside the area show that ponderosa pine forests in eastern Oregon and Washington sustained low severity burns about once a decade (Whitlock and Knox 2002), having a natural fire frequency of between 6 (Soeriaatmadja 1966) and 47 (Weaver 1959, 1961, 1967) years, depending on elevation. For western North America as a whole, ponderosa pines are believed to have burned every 2 to 25 years (Baker and Ehle 2001, Fulé et al. 2003). Fire seems to have been a frequent, recurring disturbance that kept vegetation in the region in dynamic equilibrium.

Small fires have differential effects on the plant species in the study area. Small fires tend to promote the growth of ponderosa pine, at the expense of less fire tolerant species such as spruce, firs, hemlock, and sagebrush, which attempt to invade the understory of the ponderosa pine woodlands (Martin et al. 1977, Rauw 1980). Fire tends to open the shrub steppe by killing woody species like sagebrush and promoting the resprouting of native bunchgrasses, providing a greater volume and quality of grazing materials for domestic and wild animals (Daubenmire 1970, Martin et al. 1977). Fire also causes nearly-dormant shrubs to resprout, so that more forage is available to herbivores after a fire. Berry crop production is promoted by the increased amount of available nitrogen in the soil (Dell 1977). Burning also serves to kill plant parasites (like dwarf mistletoe) (Martin et al. 1977) and reduce the fuel load on the landscape, which results in lower danger of truly large intensely destructive fires (Cairns 1980, Pyne 1982). In this area, then, fire has many uses and impacts on both the vegetation and the animals that rely on this vegetation. Important to this study is the influence that fire and fire frequency can have on human foraging economies, and the role that people have played in local fire regimes.

Area-Specific Predictions

Based on knowledge of long-term processes that have been documented elsewhere around the globe (Chapter 1) and the information presented for the study area and surrounding region, specific expectations for past events can be generated for the Rufus Woods Lake area. Native peoples were ethnographically known to burn the vegetation, resulting in the increase of grasses, decrease of sagebrush, and maintenance of open ponderosa pine forests (suppressing growth of invading, moisture-loving, arboreal species such as fir and hemlock). In other parts of the world, prehistoric anthropogenic fire is identified from increased deposition of charcoal into area lakes, with concomitant increases in early-successional plant pollen. Such disturbance signals are strongest when human population is dense or when people abruptly increase in number during episodes of colonization. In eastern Washington, one would expect there to be evidence of prehistoric burning by native people until the historic suppression of fire about 100 years ago. The charcoal record, then, should show high influxes of charcoal from 1,000 to 100 years ago, followed by a decrease at the end of the record. Since fluctuations in population size have been documented for the past thousand years, times of increased population should

correlate with or precede times of relative abundance in charcoal, and times of lowered population should correlate with or precede times of relative scarcity of charcoal in the record. If human action is driving this relationship, then charcoal should never be seen leading population in a statistically significant relationship.

Fire frequency is expected to decrease starting 100 years ago, due to the imposition of fire suppression around this time across the entire Northwest. Other consequences of the imposition of Euro-American land management practices are more difficult to anticipate. On the one hand, decreases in fire should result in the increase in trees and shrubs on the landscape. On the other hand, the introduction of large numbers of domestic herbivores should result in the relative decrease of trees and shrubs as herbaceous "disturbance" taxa such as grasses, composites, and chenopods are favored by increased grazing. It is hard to predict, given these two opposing forces, if ruderal taxa should increase or decrease at the end of the pollen record. Palynological research from other areas of eastern Washington, however, indicate that grazing pressure had the greatest effect, resulting in increases in non-arboreal pollen (especially chenopods) and decreases of all arboreal types (Davis et al. 1977).

Given the past and present vegetation, and what is known about the impact of fire in this area, specific predictions can also be made for the response of pollen to fluctuations in the population proxies. Times of greater population should either coincide with, or directly precede, increases in grasses and herbaceous members of the sunflower family (Asteraceae). Increases in grasses and other herbaceous taxa should correlate with arboreal colonizers such as alder, fir, and Douglas fir, but should be negatively associated with sagebrush and climax species such as pine. It is expected that this area will provide a clear signal of anthropogenic fire and vegetation change, and that these changes will vary directly with population size.

What of the relative effect of humans as compared to climate? Given that past research has not identified significant climatic responses in this area's vegetation over the past thousand years, climate is not predicted to be a significant factor in determining human activity, vegetation, erosion, or fire frequency.

The data generated by this study will be used to address the predictions listed above, but can also be used to address the adequacy of existing protocols and the constraints imposed by the history of the landscape system itself. If human population is positively correlated with charcoal influx in the absence of climatic change, the current practice is to deem this to be clear evidence of anthropogenic change. What of the contributions of other factors, however? What about factors such as prior fires or vegetation type? Can a simple correlation between human action and fire, in the absence of a temperature or precipitation change, be attributed to human impact? How can factors be examined, especially when intertwined?

One of the strengths of the present study is that it can statistically evaluate the relative contribution of humans in a multivariate model, while controlling for the interdependence of the human population variable with other variables. This can be done because Campbell's paleodemographic work provided a metric measure of population taken at evenly spaced intervals throughout the period of interest. The proxy population measurements, with their even temporal spacing, allow for time series analysis to be conducted. Leads and lags can be entered into statistical models comparing population with other variables like plant populations and fire. Instead of merely correlating events, relative contributions and temporal relationships can be established to more securely identify potential causal connections between the various processes operating long-term on the landscape. Currently, records from eastern Washington are anecdotal, but point to significant use of fire by hunter-gatherer populations on the Plateau. The current study can evaluate this, using quantitative means to statistically test this scenario. The Rufus Woods Lake study can also evaluate the validity of the traditional manner of using nominal and narrative information to identify anthropogenic influences. If the same conclusions are reached using time series analysis as were reached using more circumstantial evidence, this will substantiate the less rigorous methods traditionally used to identify human impacts on the ecosystem and environment. Conversely, if quantitative methods cannot identify an anthropogenic signal, this calls into question the validity of our currently popular (often qualitative) method of identifying anthropogenic change in prehistoric records.

Chapter 4. Field Work, Laboratory Work, and Preliminary Results

Field Work

Since archaeological and demographic information was already available in the study area, new work focused on the collection of comparable paleoenvironmental data. Although paleoecological information could be drawn from a number of disparate sources, lake cores were sought as a single source for all proxies to satisfy the needs of the research design. For example, existing dendroclimatological and historic records do not extend back far enough to cover the entire period of interest (Campbell 1989, Fritts and Shao 1992); tree-ring records in the Pacific Northwest typically span 500 years, with the longest reliable record reaching back only 600 years (Whitlock and Knox 2002).

Fire scar records and soil profiles provide wonderful detail, but are short in temporal duration and focused on small local or stand-level spatial areas (Arno and Sneck 1977, McBride 1983), making them too narrow, spatially, to match the human behavioral record. Historic records and repeat photography also do not extend back far enough to be of use in the current study. Archaeobotanical samples, packrat middens, and archaeofaunas all provide interesting paleoenvironmental data, but are biased by past selective collecting behaviors and are discontinuous or "patchy" in nature (Swetnam 1999). It is much more difficult to synthesize a large number of discrete depositional events from a discontinuous record into a continuous series of a thousand-year long record of 50-year intervals than it is to use a single continuous record to provide the same reconstruction. Because lake cores represent continuous accumulations of materials over the entire area and period of interest, they are simpler and easier to use in reconstructing the past. Also, lake cores have the advantage of simultaneously and continuously accumulating a wide range of materials that can be used to reconstruct many variables from a single source. By using lake sediments, past temperatures, precipitation, vegetation, salinity, past landscape fires, and erosional episodes can all be derived from a single sedimentary record (Birks and Birks 1980, Bradley 1999), ensuring that records are synchronous. This provides a secure correlation between records sampled at equal and simultaneous 50-year intervals for each proxy of past conditions over the thousand-year period of the study. This, in turn, ensures that records can be sampled at the same temporal intervals and temporal scale and therefore can be entered into a time series analysis (in the time domain) with one another (Cromwell et al. 1994, Gottman 1981, Hamilton 1994, Ostrom 1990).

Coring Site Selection

The selection of lakes for coring was guided by the need for a paleoenvironmental record that was comparable, both spatially and temporally, with the already existing demographic and archaeological records for the Chief Joseph Dam area (Campbell 1989, Linse 1993, Stein 1993). Appropriate lakes for this study would be those that primarily receive inputs from local and extralocal (i.e., about 50 to 100 square kilometers [Delcourt et al. 1983, Jacobson and Bradshaw 1981, Schoonmaker and Foster 1991]), rather than regional (i.e., over 100 square kilometer area) sources. Pollen and charcoal can be carried long distances to basins through suspension in the air, and most of the airfall pollen and charcoal that are deposited in lakes are reflective of regional vegetation (Bradshaw 1981, Jacobson and Bradshaw 1981, Faegri and Iverson 1989, Tauber 1965). Large lakes tend to have large surface areas, and hence accumulate many pollen grains from airfall onto the surface of the water. In contrast, particles enter lakes from local and extralocal sources primarily via gravity, slope wash, and stream input (Bradshaw 1981, Jacobson and Bradshaw 1981, Faegri and Iverson 1989, Tauber 1965). Small lakes with little surface area will minimize airfall input; these are the most likely to provide information on the correct spatial scale for this research.

Previous work on site selection for paleoecological applications provides guidance for finding appropriate coring locations (Jacobson and Bradshaw 1981). The contribution of local, extralocal, and regional sources is controlled by the transport mechanisms bringing particles to lakes. Fortunately, the relative input from each of these sources is correlated with the amount of surface area of a given lake, with larger lakes preferentially receiving a greater regional signal and smaller lakes preferentially receiving a local signal (Jacobson and Bradshaw 1981). To better isolate a clear extralocal signal appropriate to the scale of this study, research indicates that basins with significant local inputs from surface water (i.e., incoming streams) should be avoided. Empirical results from basins of different sizes showed that an extralocal signal could be optimized by coring a closed basin about 1 hectare in water surface area (Jacobson and Bradshaw 1981).

Lakes meeting the size and input criteria listed above were identified from USGS 7.5' quadrangle maps. Of these lakes, those that were difficult to access were excluded for logistical reasons, since these were not easily accessible for fieldwork. In 1994, about 60 potential lake sites were visited. About two-thirds of these basins were excluded from study at this time because they were currently dry, had evidence of extreme disturbance (by cattle, road construction, and farm vehicles), or because access permission could not be obtained. Ten likely candidates for coring were measured for maximum depth, and any lakes less than one m deep were also excluded from the study, since they were assumed to have been easily desiccated in the past and thus to have a high potential for a discontinuous sedimentary record. At this stage, six lake basins were identified as good candidates for coring.

These six lakes, and a nearby bog, were revisited and cored in 1997, 1998, and 1999. The basins range from 525 to 750 m in elevation and are located of both sides along the Columbia River corridor that is defined by the study area. The basins chosen for coring were Murphy Lake, Rinker's Lake, Rebecca Lake, Five Lakes, Hidden Lake, Bovy Bog, and Bedard Lake. The location of each lake is shown in Figure 3.5.

Lake Coring

Two cores were recovered from the middle each of the seven basins, where high sedimentation rates and continuous deposition was most likely (Davis et al. 1984). Coring methods varied slightly depending on equipment availability. Bedard Lake, Bovy Bog, and Rebecca Lake were cored with a 5-cm diameter, multi-stage Davis-Doyle square-rod piston corer from a wooden platform attached to an inflatable pontoon boat. Hidden Lake, Five Lakes, Murphy Lake, and Rinker's Lake were cored with a 5-cm diameter, 1.5 meter long single-stage piston corer (designed by the author) from a wooden platform attached to two inflatable rafts.

Lakes were cored to varying depths depending on the penetrability of the sediments and the nature of the coring device used. Since regional lakes were known to accrue sediments at an average rate of 1 mm per year (Forbes 1987) and the period under study covers the past 1,000 years, the goal was to recover at least 1.5 meters of sediment from each basin. Sediment recovery was recorded in the field, and cores were transported back to the University of Washington in Seattle for description, storage, sampling, and laboratory analysis. Of these six cores, three (Murphy Lake, Rebecca Lake, and Bovy Bog) had records that were either too short for analysis, or showed discontinuities that made them ineligible for further analysis (see Scharf [2002] for details). This left Hidden Lake, Rinker's Lake, and Five Lakes for further study. Their locations are provided in more detail in Appendix B (which lists the latitude, longitude, UTM, and elevation of each), and each is described briefly, below.

The local setting of Hidden Lake is detailed in Figure 4.1 and shows that this lake has no input or outflow stream and sits within a local depression between "fins" or linear ridges of bedrock in the highlands above Goose Flats. Given its geographic context, Hidden Lake can be expected to gain most of its input from local sources. This particular basin is located near stands of ponderosa pine, patches of sagebrush-steppe, and locally wet and sheltered areas that harbor small stands of deciduous trees. This lake basin lies about 2 miles (~3.2 km) north of the Columbia River channel.

Rinker's Lake (Figure 4.2) lies to the south, on the other side of the Columbia River. Rinker's Lake is a kettle hole left in the glacial outwash, and is located in a small south-draining depression in the grass-sagebrush steppe in Douglas County. This basin, like Hidden Lake, is small, has no input or outflow stream, and sits within a small depression in the landscape, and is therefore likely to be receiving primarily local pollen inputs.

Finally, Five Lakes is located in the highlands east of the Columbia River, south of the modern town of Nespelem and approximately one mile (~1.6 km) northwest of Buffalo Lake in Okanogan County. Five Lakes is a series of small water-filled basins that lie within a depression in the highlands east of the Columbia River, a little over four miles (~6.4 km) northeast of Buckley Bar. These basins, likewise, are small and have neither inflow nor outflow streams. Again, these are likely to accrue sediments (including pollen) from local sources. The map in Figure 4.3 shows this setting, specifies which basin was cored, and the specific location within that basin that was used for coring. For this basin, as well as Rinker's and Hidden Lakes, sediment was taken for analysis via coring with a piston corer at the deepest portion of the lake.

Laboratory Protocol

Initial Processing

Within a day of collection, cores were transported to a laboratory facility where the color and lithology were described, and cores were put into refrigerated storage before further study. Unconsolidated sediments were extracted at 0.5 cm intervals and placed into labeled re-sealable plastic bags. At this stage, the location and nature of stratigraphic boundaries were recorded, and note was made of any visual or textural evidence for discontinuities, unconformities, or other anomalies. Sediment descriptions were completed for the Hidden, Rinker's, and Five Lakes cores, showing that each consisted of a single sedimentary layer with no evidence of interruption or change in sedimentation.

Loss-on-Ignition Analysis

After basic sediment description, lake cores from all basins were sub-sampled for loss-on-ignition (LOI) analysis. LOI analysis basically followed standard procedures as described in Dean (1974) and the steps are outlined in Appendix C.1. As Heiri and others (2000) point out, variability in results can be due to sample composition or from analytical error. Since bias can be introduced by varying sample size and duration of burns, standard procedures (equal temperature and dwell periods) were used throughout this study to ensure consistency and comparability between samples. Obtained from this analysis (Dean 1974, Konrad et al. 1970) was an estimate of the percent composition, by weight, of organic materials (such as lignin and cellulose) and carbonates (probably mostly aragonite support structures from aquatic plants and the shells of ostracods). LOI results were scanned to identify potential discontinuities or rapid changes in sediment composition. Results from the microcombustion analysis were also used to determine which samples were most appropriate for pollen research and which were better suited to oxygen isotope analysis.

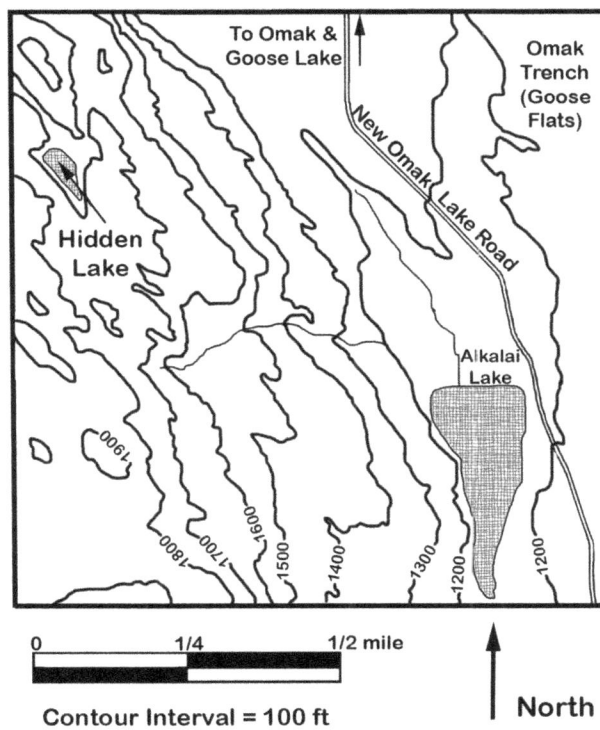

Figure 4.1. Detailed Location Map of Hidden Lake
(Adapted from USGS 7.5' Map—Boot Mountain, WA 1980)

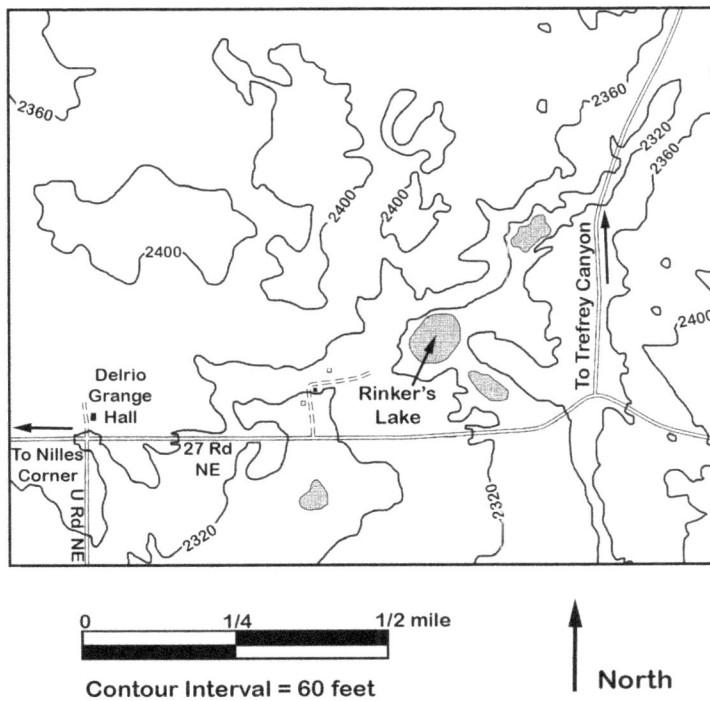

Figure 4.2 Detailed Location Map of Rinker's Lake
(Adapted from USGS 7.5' Map—Trefry Canyon, WA 1980)

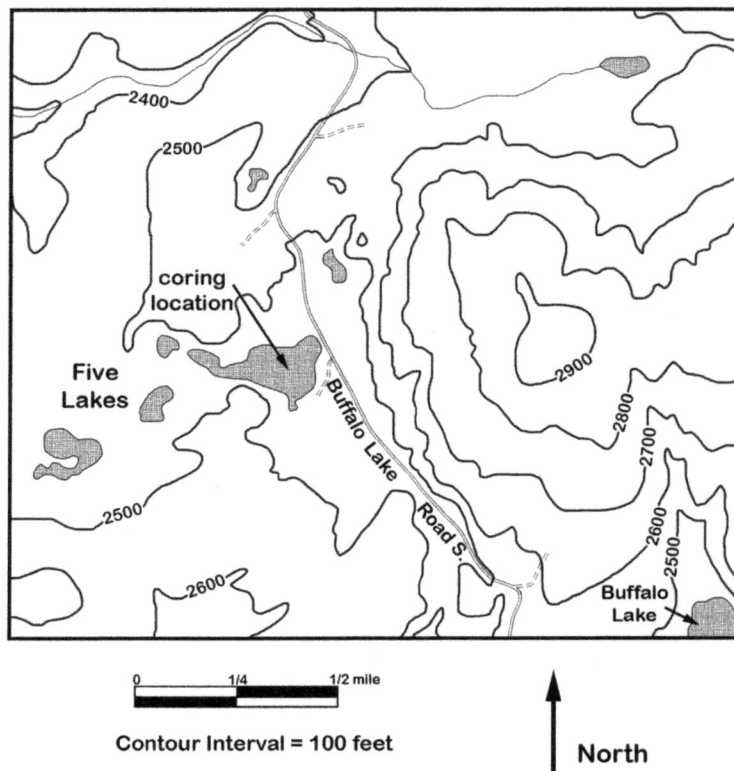

Figure 4.3 Detailed Location Map of Five Lakes Coring Site
(Adapted from USGS 7.5' Map—Belvedere, WA 1989, Provisional Edition, Revised 1993)

Radiocarbon Dating

Bulk sediment was submitted for radiocarbon dates. Each radiocarbon sample was restricted to a span of about 0.5 cm of the core, in an attempt to restrict the span of time represented by the deposits. Accelerator Mass Spectrometry (AMS) radiocarbon dating was chosen since the resulting samples were small (each about 2 to 4 g) and because AMS, as a direct method of determining radiocarbon content, yields results with narrower error estimations than conventional radiocarbon dating techniques (Brown et al. 1989). Samples from the basins were submitted to the University of Arizona laboratory for AMS dating. Dates were corrected for isotopic fractionation by the laboratory and calibrated by the author using CALIB 4.3 (Stuiver and Reimer 2000; an update to Stuiver and Reimer 1993). Where possible, calibrated dates (Taylor 1987) were converted to dates B.P. (Before Present).

Along with the LOI analysis, AMS dates provided a means for assessing the sedimentation rates and changing sedimentary regimes in each lake basin. At this stage, potential stratigraphic problems like mixing, reversals, and hiatuses were identified in some of the lakes, leaving only certain cores fit for further study. Of the cores with no visible sedimentary problems, AMS dates provided chronological controls for sub-sampling the sediment cores for further analysis. In the initial stages of the study, LOI and visual examination of sediments under a binocular dissection microscope were used to attempt to locate and identify possible dateable tephra layers to be used for additional chronological controls. Volcanic ashes from the last 1,000 years in eastern Washington were small events, are often not preserved, and are thin stratigraphic markers where they do exist. The lakes chosen for study had relatively small catchments and had not concentrated ash from the surrounding landscape in a way that would make the ash layers visible or easily identified. Tephrochronology efforts were therefore dropped from the analysis at this stage.

Since sedimentation rates could be estimated from AMS dates, samples could be taken at approximately 50-year intervals from the cores for pollen and charcoal analysis. This strategy was followed so that the proxy lake records produced would have the same temporal resolution as the already-established demographic record for the study area. This sub-sampling strategy, along with lake-site selection, promised to deliver proxy records that were fully comparable to Campbell's population reconstruction (Campbell 1989, 1990).

Preliminary Results: Suitability of Basins and Cores for Further Study

Lakes were chosen for coring based on their morphology, hydrology, and location, in order to maximize an extralocal spatial-scale signal appropriate for the study of human activity. After coring, lake and bog sediments had to meet four additional criteria to be used as a source of proxy records for this research. Each basin had to provide a record that (1) was the product of continuous sedimentation (not the product of a series of discrete depositional events), (2) reflected changes in the sources of particles and not changes in sedimentary processes transporting and depositing the particles, (3) spanned the entire period of interest from the present back to A.D. 950 (1,000 B.P.), and (4) contained materials appropriate for the analyses being conducted.

The rest of this chapter evaluates the suitability of each core, in turn, using as evaluative criteria a combination of radiocarbon dating, sediment descriptions, and LOI analysis. The results of these techniques indicate that Hidden Lake, Rinker's Lake, and Five Lakes were appropriate for further study. Of these lakes, Five Lakes was most likely to provide a solid and well-preserved pollen record, while Hidden and Rinker's Lakes were better suited for oxygen isotope analysis.

Evaluating Cores for Continuous Deposition

Evaluating interactions between people, vegetation, climate, and fire logically requires complete records for each variable, without temporal discontinuities. The use of time series and other statistical analyses, likewise, demands that there be no temporal gaps or missing values in the data set being used. An uninterrupted temporal sequence is necessary given both the questions being asked and the statistical techniques being used.

A continuous record is a traditional requirement of palynological studies, and simple techniques have been established for evaluating cores on this basis (Faegri and Iversen 1989). Palynologists commonly use radiocarbon dating to estimate the sedimentation rate and combine this information with sediment descriptions to identify problems in lake cores (Faegri and Iversen 1989, Webb and Webb 1988). The apparent sedimentation rate in a lake core is calculated by using linear interpolation between two dates from different stratigraphic depths (Webb and Webb 1988). Normal sedimentation rates for lakes average 0.91 millimeters per year (or mm/yr), with an associated 95% confidence interval from 0.16 to 2.57 (Webb and Webb 1988). An abnormal sedimentation rate lower than 0.16 mm/yr is "characteristic of nonconstant processes of accumulation" (Webb and Webb 1988: 293) and often signals the presence of a hiatus in deposition or net erosion from a basin. An apparent mean sedimentation rate can be calculated for any lake sediment interpolating between samples of known date and stratigraphic depth. This is done by simply dividing the stratigraphic distance (in mm) between two samples by their age difference. For example, a modern core top date and a date of 500 B.P on a sample from 1,000 mm b.i. (below interface—the stratigraphic depth with regard to the sediment-water interface) would be associated with an apparent sedimentation rate of 2 mm/year as (1,000-0 mm)/(500–0 yrs) = 2 mm/yr.

Previous work in the study area itself has shown that these apparent sedimentation rates are useful guides for distinguishing between disrupted and uninterrupted sedimentary sequences. In three separate analyses of Goose Lake, researchers found no evidence for discontinuities in the core (Dalan 1985a, Forbes 1987, Nickmann and Leopold 1985). Radiocarbon dates from Goose Lake (from Forbes 1987), when calibrated, confirm this (Table 4.1). The Holocene sedimentation rates at Goose Lake varied from 0.43 to 1.11 mm/yr, in agreement within expectations for a conformable sequence. Earlier work on Duley Lake, in contrast, showed that the upper part of the core contained an unconformity (Forbes 1987). Not surprisingly, the associated sedimentation rate calculated from calibrated radiocarbon assays at Duley Lake was anomalous at 0.03 mm/yr (Table 4.1). An analysis of sedimentation rates, then, promises to provide a good method for screening additional basins for potential stratigraphic problems.

In order to evaluate the basins cored for the present study, bulk sediment samples were taken from each core and submitted for AMS radiocarbon dating. The samples submitted, along with the results in ^{14}C years (uncalibrated radiocarbon years before present corrected for δ^{13}C), are listed in Table 4.2. Radiocarbon dates were calibrated using Calib v.4.3 calibration program (Stuiver and Reimer 1993), and results are listed in Table 4.3, along with the 2σ (95.4%) confidence interval for each calibrated date. In addition to AMS dates, chronological control was also provided by the core-top as well. Since deposition was occurring at the sediment-water interface at the time the cores were collected, 0 mm b.i. was assumed to have a modern date.

Along with the mean calibrated radiocarbon dates and core-top stratigraphic dates, linear interpolation was used to calculate apparent sedimentation rates. The results of these calculations are presented in Table 4.3 and will be discussed on a case-by-case basis. In addition to sedimentation rate, sediment descriptions were used to identify potential interruptions in sedimentary sequences. This information is presented for each basin in Figures 4.4 through 4.6.

Evaluating Cores for Uniform Transport and Deposition

Although it is important for basins to be constantly receiving sediment, they must also be receiving material without undergoing significant changes in the mechanisms delivering the sediments. This is important in the analysis because measured changes in each proxy variable need to reflect changes in the underlying variable of interest rather than changes in depositional processes. Changes in sediment sources, transport agents, or the environment of deposition can all interfere with proxy signals. For example, increases in the amount of grass pollen in lake sediment samples cannot be assumed to reflect changing abundances of grass on the landscape if there have been fluctuations in the amount of sheetwash transporting pollen into the basin or a change in preservation of pollen over time.

Table 4.1. Holocene Dates and Sedimentation Rates from Previous Work on Goose and Duley Lakes

Lake Name	Depth in Core (mm b.i.)	[14]C Date ^{14}C yrs	Mean Date(s) cal B.P.	95.4% C.I. (2σ) for Mean Date	Apparent Sedimentation Rate (mm/cal yr)	95% C.I. for Apparent Sedimentation Rate	
						Slowest	Fastest
Goose Lake	600	1,420±20	1,307	(1,290-1,348)	0.45	0.43	0.45
	4,100	4,850±60	5,593	(5,688-5,707)	0.82	0.80	0.84
				(5,672-5,680)			
				(5,566-5,661)			
				(5,470-5,559)			
	6,300	6,770±50	7,611	(7,566-7,682)	1.10	1.08	1.11
			7,595	(7,511-7,529)			
			7,593				
Duley Lake	100	2,930±30	3,135	(3,180-3,208)	0.03	0.03	0.03
			3,132	(2,958-3,167)			
			3,077				
	1,600	8, 170±70	9,126	(9,377-9,399)	0.25	0.17	0.25
			9,105	(9,357-9,373)			
			9,088	(9,341-9,350)			
			9,045	(9,310-9,325)			
			9,033	(8,999-9,303)			

Table 4.2. AMS Radiocarbon Dates for Lakes Cored for This Study

Lake Name	Depth in Core (cm b.i.)	Weight of Sample (g)	Lab Number	Material Dated	[14]C Date ^{14}C yrs
Hidden Lake	25	2.0	AA34828	dry lake sediment (12% organic)	500±50
	50	2.3	AA34829	dry lake sediment (10% organic)	Not reported by ^{14}C lab
Rinker's Lake	53	2.9	AA34948	dry lake sediment	1,660±55
Five Lakes	50	3.6	AA34830	dry lake sediment (40% organic)	580±45
	75	3.6	AA34831	dry lake sediment (31% organic)	820±45
	100	3.9	AA34832	dry lake sediment (40% organic)	1,050±55
	125	3.9	AA34833	dry lake sediment (46% organic)	1,325±45

Table 4.3. Dates and Sedimentation Rates from Lakes Cored for This Study

Lake Name	Depth in Core (mm b.i.)	^{14}C Date ^{14}C yrs	Mean Date(s) cal B.P.	95.4% C.I. (2σ) for Mean Date	Apparent Sedimentation Rate (mm/cal yr)	95% C.I. for Apparent Sedimentation Rate	
						slowest	fastest
Hidden Lake	250	500±50	524	(602-624)	0.44	0.37	0.48
				(477-558)			
Rinker's Lake	530	1,660±55	1,543	(1,413-1,705)	0.33	0.30	0.36
Five Lakes	500	580±45	593	(518-653)	0.78	0.71	0.88
	750	820±45	729	(870-881)	1.68	1.09	1.89
				(814-826)			
				(667-792)			
	1,000	1,050±55	953	(909-1,061)	1.10	1.39	1.91
				(833-854)			
				(798-809)			
	1,250	1,325±45	1,271	(1,172-1,307)	0.79	0.67	1.02

Evidence for steady versus changing sedimentary regimes can be garnered from several lines of evidence. Sedimentation rates, again, help identify problematic sequences, as radical changes in accumulation rates signal gross changes in sediment transport and/or origins. Sediment descriptions can be used to identify any variation(s) in lithology that may be reflective of changes in depositional processes as well. Additionally, LOI can be used to trace trends in the relative contribution of different sediment sources by quantifying the organic, carbonate, and other inputs into lakes. Marked changes in LOI values identify shifts in accumulation of organic or other materials into a basin signal a change in depositional environment and/or transport mechanisms into the basin (e.g., Nesje and Dahl 2001), all of which affect pollen transportation, deposition, and preservation.

Evaluating Cores for Record Length

Even steady and continuous sedimentary records will be useless for study unless they extend back to A.D. 950. Some of the cores have AMS dates drawn from the lowest stratigraphic depth in the sequence, and this is a direct indication of the time-span represented by the core. For other cores, the lowest stratigraphic depth is below the lowest dated material. In this case, sedimentary rates and AMS dates are used to extrapolate downcore to calculate a date for the oldest material recovered. For example, the Five Lakes core is 1,450 mm long, but the lowest date of 1,271 B.P. comes from 1,250 mm b.i. Extrapolating, using the mean apparent sedimentation

rate of 0.79 mm/yr (from Table 4.3) associated with the two lowest dates, the additional 200 mm of sediment should represent an additional 254 years of deposition for a total of 1,525 years of sedimentation in the core. Record lengths as determined from radiocarbon dates and sedimentation rates are presented in Table 4.4. As this table shows, all three basins cover the entire period of interest.

Evaluating Cores for Appropriate Material Types

Also of interest is the chemical and particulate composition of the cores, which determines whether they are appropriate for the laboratory techniques employed to measure the proxy variables. The most useful pollen record would be derived from a basin that originally received a high, steady input of pollen and had good preservation of that pollen. Basins to be avoided would be those with relatively low organic content or with evidence of shallow or dry conditions. Promising basins would have the opposite characteristics—high relative organic composition, acidic conditions, and lithology consistent with open water conditions. Gyttja, a highly organic lake mud, indicates open water conditions favoring pollen accumulation and preservation. On the other hand, sediments with high inputs of terrigenous sediments (like sand) often point to unfavorable emergent or drying conditions in which little pollen is deposited and destruction of grains is high. In this study, sediment descriptions and LOI results are used to identify those

Table 4.4. Temporal Length of Sediment Records Extrapolated from Calibrated AMS Dates

Lake Name	Column A Core Length (mm)	Column B Lowest ^{14}C Date (mm b.i.)	Column C Oldest Date (cal B.P.)	Column D Mean Sedimentation Rate (mm/yr)	Column E Record Length (yrs)**	Column F Oldest (Lowest) Sediment
Hidden Lake	560	250	524	0.44	1,235	cal A.D. 715
Rinker's Lake	530	530	1,543	0.33	1,543	cal A.D. 407
Five Lakes	1,450	1,250	1,271	0.79	1,525	cal A.D. 425

*all dates and years, unless otherwise noted, are calibrated

**Column E, the record length, was determined in the following manner:
If the stratigraphically lowest sample was dated, the record length (Column E) simply equals the lowest calibrated date (Column B). Otherwise, the age of the lowest sample was calculated by extrapolating from the lowest known date to the lowest point in the core, using the average calibrated sedimentation rate. This can be expressed for the columns as E=((A-B)/D)+C

cores with highly organic lake sediments that are reflective of open water conditions.

Lithology and LOI can, likewise, be used to identify sediments with compositions likely to preserve a strong, clear oxygen isotope record that is reflective of changing climate. Since oxygen isotope assays are done on the carbonates in lake sediments, LOI measurements can identify those sediments with high concentrations of the necessary minerals (i.e., carbonates). Lithologic descriptions can, again, be used to differentiate between sediment sequences derived from basins that underwent emergent periods and those that permanently maintained open water. This is necessary because one of the biggest potential confounding factors in oxygen isotope analysis is changing water volumes in lakes. Promising candidates for isotope work are marls (Forbes 1987), which are carbonate-rich and usually accumulate in standing water. Especially desirable would be sequences of uninterrupted marls, indicative of constant open water conditions and carbonate deposition over long periods of times. Much less desirable would be sediments low in carbonate, or those indicative of fluctuating water depth and chemistry, such as alternating bands of peat and marl.

Basin-specific Preliminary Results and Conclusions

Hidden Lake

Hidden Lake produced a 56 cm-long core with a date (from 25 cm b.i.) of 524 years. The apparent mean sedimentation rate of 0.44 mm/yr for Hidden Lake (Table 4.3) is consistent with both the expectations of "normal" lacustrine sedimentation (Webb and Webb 1988) and similar to the average rate of 0.45 mm/yr seen in the upper portion of the core from nearby Goose Lake. The

stratigraphy from Hidden Lake (Figure 4.4) is simple and consists of a single unit of silty marl indicative of constant and continuous deposition in open water. Based on the depth of the calibrated AMS date, the Hidden Lake sediments represent an estimated 1,235 year sequence dating back to A.D. 715, providing a record that is adequate for the research questions under study.

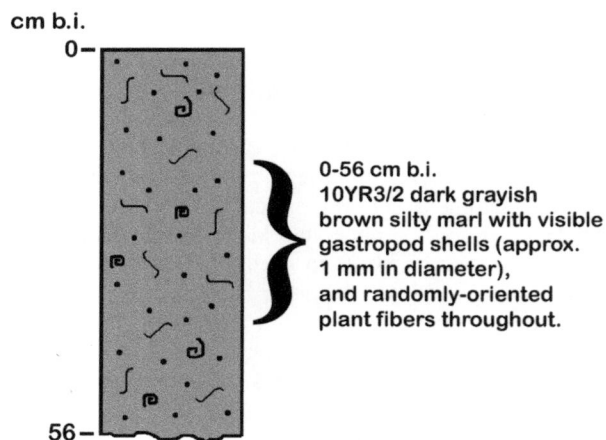

0-56 cm b.i.
10YR3/2 dark grayish brown silty marl with visible gastropod shells (approx. 1 mm in diameter), and randomly-oriented plant fibers throughout.

Figure 4.4. Sediment Description for Hidden Lake

Rinker's Lake

Rinker's Lake is also promising, and the sediment description for this record is presented in Figure 4.5. Rinker's Lake sediments consist of a 53 cm-long unit of silty clayey marl. This unbroken layer of marl indicates that the basin has steadily accrued material under open-water conditions. This is consistent with an apparent sedimentation rate of 0.33 mm/yr (Table 4.3) calculated for these sediments, and falls within the 95% confidence interval (0.16-2.57 mm/yr) considered normal for North American basins (Webb and Webb 1988). Rinker's Lake, then, fits the requirements of uninterrupted and unchanging deposition. With a calibrated date of 1,543 years (A.D. 407—Table 4.4) from the core bottom, this basin satisfies the temporal length requirement as well.

cm b.i.

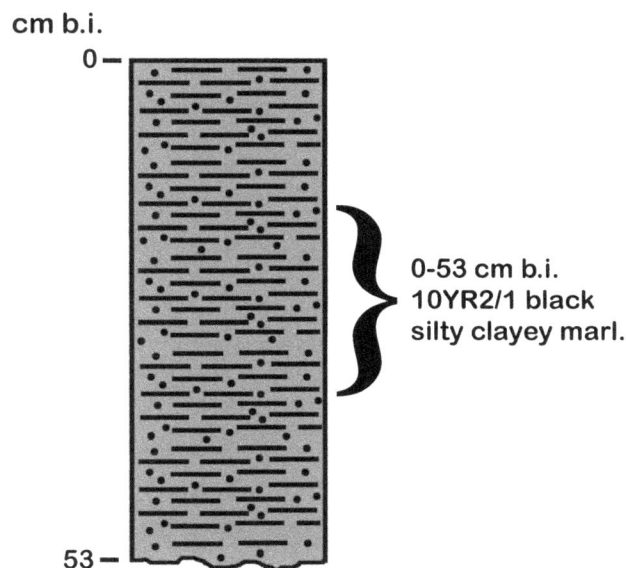

Figure 4.5. Sediment Description for Rinker's Lake

Five Lakes

Five Lakes also satisfies the requirements for continued laboratory analysis. This lake was the source of a 145 cm-long core composed primarily of gyttja, and this record shows no major changes in lithology, nor any other evidence of drying conditions (Figure 4.6). Four samples from this sequence were submitted for radiocarbon dating, and the series (see Tables 4.2 and 4.3) of dates conforms to all expectations—radiocarbon ages increase with depth, sedimentation rates do not vary significantly throughout the core, and the apparent sedimentation rates, which range from 0.78 to 1.68 mm/yr, are well within the normal range. Again, this basin shows strong evidence that it has an unbroken and unchanging depositional history. The length of this history can be extrapolated from the stratigraphically lowest AMS date and its related sedimentation rate. The results show that the Five Lakes core spans 1,525 years (back to A.D. 425), more than the minimum requisite. Additionally, the presence of gyttja throughout the core indicates that this basin contained standing water throughout its past 1,525 years.

Basins Selected for Pollen and Isotope Work

Five Lakes can be compared with Rinker's Lake and Hidden Lake to determine which sediments are most appropriate for oxygen isotope analysis and which are best suited for pollen and charcoal work. Of the three

basins, the gyttja from Five Lakes has the highest organic content, with an average of 31.2% organics by weight. The marls from Rinker's and Hidden Lakes, in contrast, have only 5.2 and 20.8% organics, respectively. Based on organic content, Five Lakes can be expected to have the best pollen record and was chosen as the source of the vegetation and fire proxies.

Rinker's and Hidden Lakes, in contrast to Five Lakes, are composed of lake marls with relatively high (15.8 to 23.6%) carbonate content. Previous work with oxygen isotopes from other area lakes has confirmed that marls are the best choice for $\delta^{18}O$ work, since they give the clearest climate signal, the most consistent results and the best sample sizes (Forbes 1987). Based on their lithology and LOI, Rinker's and Hidden Lakes were determined to be more suitable for oxygen isotope analysis.

cm b.i.

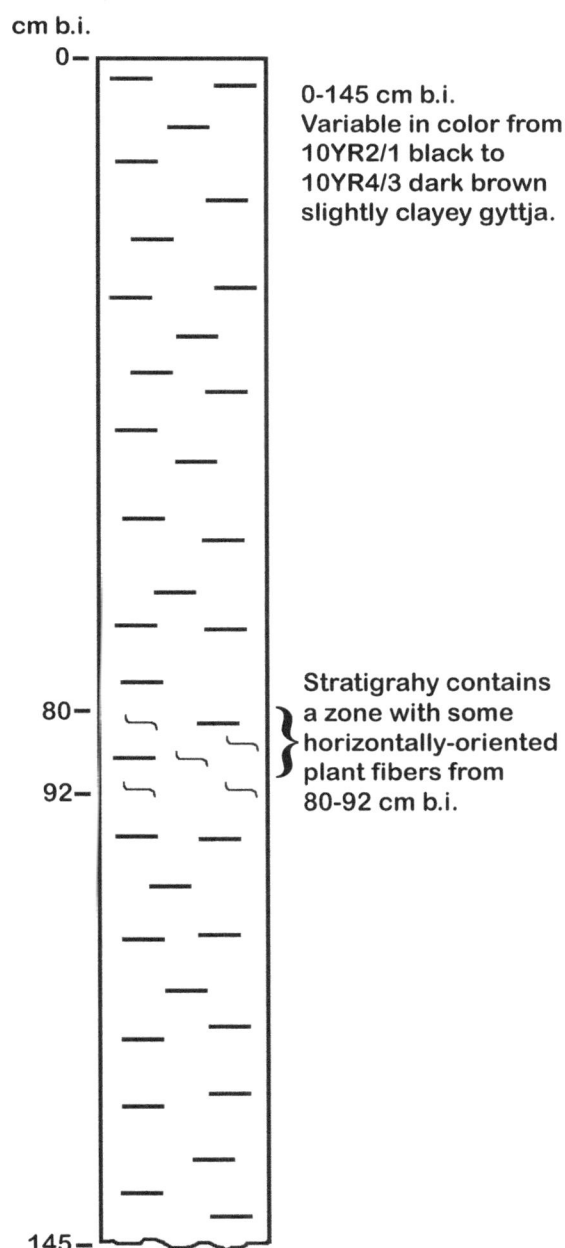

Figure 4.6. Sediment Description for Five Lakes

Pollen Analysis

The vegetation proxy for this study was constructed using pollen from sediment samples. Standard chemical processing procedures were followed to digest and eliminate organic and inorganic materials found in the sediment, while retaining and concentrating pollen grains (Appendix C.2, Faegri and Iversen 1989, Gray 1965).

After processing, pollen samples were examined under a 400x stereoscopic light-transmissive microscope according to established procedures and conventions (Faegri and Iverson 1989). Using transects that covered the entire visible field of the slide, pollen grains were identified to the lowest possible taxon and recorded, following traditional identification methods, aided by pollen keys and reference collections (Faegri and Iversen 1989, Kapp 1969, Moore et al. 1991). Introduced *Lycopodium* spores were identified and counted when encountered along the pollen transects. An aliquot of these *Lycopodium* spores had been introduced into samples earlier to serve as a standard on which to calculate the pollen accumulation rate (PAR). These two sets of observations provided the raw information needed to create pollen diagrams for the basins. The taxa present, their percent abundance in the assemblage, their concentration, and their PAR could all be calculated from the information gathered.

Along with these traditional pollen counts, pollen condition was also noted. Counts of degraded, corroded, or otherwise damaged fossil pollen grains are an ancillary line of evidence pointing to changing sedimentary regimes and erosion since damaged pollen grains are produced by dry or alternating wet/dry conditions (Cushing 1967). Damaged pollen grains found in a lake indicate either that the lake was formerly dry, underwent radical shifts in pH, or received pollen grains reworked and transported from eroding uplands by surface water. A sudden influx of damaged grains into a wet basin could indicate the sudden arrival of eroded upland materials, which could be caused by any number of external events such as climatic change, overgrazing, deforestation, or similar disturbances.

Along with pollen, other small, chemically resistant materials were recorded when encountered during the examination of slides. Algae (mostly *Pediastrum*) were recorded when observed, as these are good indicators of past water depths (Bradley 1999) and are routinely used as monitors of past climatic conditions (e.g., Herzschuh et al. 2005, Sarmaja-Korjonen et al. 2006). Spores were also recorded as encountered, since the presence of certain spores provide significant paleoenvironmental information. For example, the presence of *Sporormiella* in lake sediments has been linked with intense herbivore grazing and high herbivore populations (Davis and Shafer 2006) during prehistoric times in the Holocene (e.g., Burney et al. 2003) and Pleistocene (e.g., Davis 1987). Abundances of this spore could track potential impacts of prehistoric hunting pressure and has already been used successfully in eastern Washington to palynologically trace the introduction of domestic herbivores in historic times (Davis et al. 1977).

Charcoal Analysis

For ease of analysis and to guarantee the synchroneity of the vegetation and fire proxies, charcoal particles greater than 75 μm were counted when encountered on the pollen slides, using methods similar to those employed by Backman (1984), Gajewski et al. (1985), and Swain (1973, 1978). Charcoal grains were counted, taxonomically identified, when possible, and measured for size (average diameter). Both counts and areas (calculated based on the average diameter) were used to create charcoal influx and grain size records, following established conventions for charcoal studies (Clark 1988b, 1988c; Maher 1972; Patterson et al. 1987).

While the preparation, counting, and presentation techniques of pollen analysis are standard, the methods used for charcoal analysis vary from researcher to researcher. Some researchers process lake sediments for charcoal by indurating the sediments with a stiff epoxy, creating thin sections, and point counting charcoal under a binocular dissecting scope (Clark 1988a, 1988b; Clark and Royall 1995b). Some (e.g., Mehringer et al. 1977a, Millspaugh and Whitlock 1995) wet-sieve lake sediment through a set of nested geological screens and count the entire number of recovered particles under a microscope. The thin section and sieving methods are often used by those seeking to count large charcoal grains to obtain local and/or lake catchment fire reconstructions. Others use less popular methods such as nitric acid digestion (Winkler 1985b), or computer image analysis (Patterson and Backman 1986). Still others (Backman 1984; Gajewski et al. 1985; Swain 1973, 1978) count a subsample of charcoal particles that are visible on pollen slides during pollen counting, and estimate the total number of particles with the aid of a reference aliquot. This last method is usually associated with the counting of small charcoal particles with the goal or reconstructing regional-scale fire histories.

Even given the dizzying array of choices available for producing a charcoal record, the choice for this study was easy—charcoal was counted using the pollen-slide method. This method was chosen for counting instead of thin sections or sieving for many reasons. First, the pollen slides could be easily sub-sampled because they contained a known amount of a standard *Lycopodium* aliquot. In previous sieving analyses, researchers noted the inconvenience or impossibility of counting the entire sample of recovered charcoal (Mehringer et al. 1977a, Millspaugh and Whitlock 1995). Second, the use of pollen slides ensured comparability between the pollen and charcoal samples. Both charcoal and pollen samples were drawn from the same original sediment sample and both received the same processing treatments, thereby preventing any bias due to differences in stratigraphic origin or laboratory treatment.

Clark (1984) has cautioned that pollen processing methods could bias charcoal results. To control for this,

Cwynar (1987) suggests that all samples be treated identically during preparation. More recently, Clark and Royall (1995b) compared charcoal results from the thin-section and pollen-slide methods. They found that results from the two methods provided agreed well with each other for prehistoric samples (Clark and Royall 1995b). The pollen-slide method is considered reliable when used cautiously, and it has many advantages over thin-section preparation and analysis. Another critique of the pollen slide method is that it can suffer from an inflation of counts if processing fragments existing charcoal grains into a larger number of small particles (Whitlock and Larsen 2001, Weng 2005). Even if this were true, equal treatment of samples during processing should result in an equivalent bias for all samples, and since Five Lakes results are only being compared to the Five Lakes dataset, this provides no analytical barrier to the present study. This would only be problematic if Five Lakes samples were inconsistently treated, or being compared to a sequence obtained by other methods, such as sieving. Additionally, the current study focuses on the macroscopic particles (125-500 μm), which should control for breakage. Logically, any breakage should result in the reduction of one large particle into several smaller ones, creating an inflated number of small particles. As such processes should not result in increased numbers of large particles, the use of only large particles should correct for potential problems.

Initial processing of thin section slides from an area lake was yet another reason for choosing the pollen-slide method. Undisturbed, unprocessed lake sediments from Bedard Lake (a lake located west of Hidden Lake) were indurated with low viscosity (SEM-grade) epoxy, cut, mounted, and polished. The thin sections thus created were examined under 40x so that charcoal particles could be counted and their diameters measured. This process proved very time intensive, and the smaller charcoal particles (which are the most abundant and ubiquitous) escaped detection using this method. As a result, the thin-section method was abandoned, and pollen-slide counting provided the charcoal record.

Just as preparation methods are variable from author to author, quantification of results is equally variable, but usually involves the calculation of charcoal influx. This is referred to as "CHAR" in the literature and is often calculated in a manner similar to that reported by Maher (1981). Some (Mehringer et al. 1977a, Millspaugh and Whitlock 1995, Whitlock and Millspaugh 1996) present data based on simple counts as $\#/cm^2/yr$ (total number of charcoal particles deposited per square centimeter per year). Others (Clark 1988b, Maher 1981) present the total area of charcoal particles deposited per year. Area estimations are often expressed as $mm^2/cm^2/year$ (Clark and Royall 1995a, 1995b), and sometimes as $cm^2/cm^2/year$ (Clark 1987, Clark 1988b) or $\mu m^2/cm^2/year$ (Cwynar 1987, Maher 1981). Also seen are presentations of charcoal data as the ratio of charcoal particles to pollen particles within a given sample (Cwynar 1978; Patterson

and Backman 1988; Patterson et al. 1987; Swain 1973, 1978), the percent of charcoal by weight in dry sediment (Winkler 1985a, 1985b) or less commonly as counts of charcoal in a given weight or volume of sediment (Fall 1997, Kirch and Ellison 1994, Umbanhowar 1996).

Charcoal on pollen slides was counted in the same manner that pollen was counted. That is, the same transects on the same slides were scanned and tallies were kept of the number of introduced *Lycopodium* (the introduced aliquot standard) and charcoal grains. Slides were counted until the 500th identifiable grain of pollen was found, and then counting continued to the end of the transect containing the 500th grain. Size information for charcoal particles was kept by recording grains under one of three size categories: 75-125 μm diameter, 125-250 μm diameter, and 250-500 μm diameter. All charcoal smaller than 75 μm was ignored, since most research has shown that particles of these sizes provide little to no information for reconstructing extralocal or catchment fire histories (Clark 1995, 1988a; Mehringer et al. 1977a; Millspaugh and Whitlock 1995; Whitlock and Millspaugh 1996; Patterson and Backman 1988). This counting method allowed both size and ubiquity of particles to be taken into consideration so that charcoal counts could be easily converted into any of the popular measures (whether size- or abundance-based) commonly used for reporting charcoal results. Charcoal quantification, along with pollen identification and recording of spores were all done by the author, while specialized isotope determinations were outsourced.

Isotope Analysis

A second set of sediment samples was used to create a climate proxy. Two indicators of past climates easily reconstructed from lake sediments are water depth and salinity, and many lines of potential evidence are available for measuring these variables (Bradley 1999). For this study, isotopic fractionation was used as the climate proxy since previous research on lacustrine sediments from Goose and Duley Lakes (see Figure 3.5 and Appendix B for lake locations) showed that isotope values observed in these fresh water lakes could and did accurately and sensitively track the area climate (Forbes 1987). More recently, other researchers in the greater interior region have used ground water isotopic composition as a means to reconstruct paleoclimates (Schlegel et al. 2009).

Sediment samples for oxygen isotope analysis and carbon isotope analysis were drawn from two lakes—Rinker's Lake and Hidden Lake. Samples were outsourced for carbon isotope analysis to Mountain Mass Spectrometry, an independent consulting firm in Evergreen, Colorado. Differences between the two records were treated as local signals that were unique to each lake, and commonalities between the two records were considered to reflect an overarching extralocal climate signal that was being manifested simultaneously in both lakes.

Chapter 5. Isotope Record (Climate Proxy) from Hidden and Rinker's Lakes

Oxygen Isotope Composition as Paleoclimate Proxy

Because oxygen isotopes vary in the number of neutrons they have, isotopes have different thermodynamic and kinetic properties. Since oxygen isotope fractionation is highly temperature dependent, the relative proportions of ^{18}O and ^{16}O are good indictors of past climatic conditions. ^{18}O has two more neutrons than ^{16}O, and this mass difference causes ^{18}O to have a lower vapor pressure than ^{16}O. Because of this, evaporation tends to enrich the remaining liquid in ^{18}O, while condensation tends to lead to greater relative amounts of ^{16}O in vapor. This is used to reconstruct records of continental ice volume changes, as this process results in a distillation of water vapor creating precipitation that becomes progressively more deleted in ^{18}O with latitude. When high latitude precipitation gets locked onto the land as snow and ice during glacial periods, the composition of ocean water becomes more ^{18}O-enriched and ^{16}O concentrations in ocean water drop (Bradley 1999).

Most past research on using $\delta^{18}O$ in climate reconstructions has focused on global-scale temperatures and ice volumes, employing marine carbonates from vast ocean reservoirs (e.g., Emiliani 1955, Shackleton and Opdyke 1973). In contrast, smaller reservoirs are more responsive to the conditions near them; terrestrial lakes are "sensitive recorders of relatively minor and localized climatic changes" (Forbes 1987:1). Isotopic composition of water in small closed lakes is dependent primarily on the balance between water coming into the lake relative to the water lost through evaporation. As a result, the relative proportions of ^{18}O and ^{16}O in water in a small closed basin are a function of the temperature and duration of evaporation. Stable isotope ratios records can therefore be used as a source of information on past climatic conditions. Oxygen isotope composition derived from marls and carbonate-producing taxa (bivalves, gastropods, plants, algae, plankton) all record water composition at the time of crystallization (Forbes 1987, Leng et al. 2001, Noon et al. 2003, Tevesz et al. 1996, Wu et al. 2007). Taking local hydrological conditions into consideration, $\delta^{18}O$ from lake carbonates is used to track hydrological and climatological changes over time in the local or regional area, often tracing trends that are not documented by global-scale reconstructions. In areas with seasonal ice and snow accumulation, oxygen isotope composition is often greatly influenced by changing meltwater inputs (Leng et al. 2001, Noon et al. 2003). Under other conditions, $\delta^{18}O$ is correlated with summer water temperatures (Leng et al. 2001, Wu et al. 2007). Paleoecologists in arid or semi-arid areas have found that small, closed lakes should undergo changes in water composition from volumetric fluctuations due to alterations of the water balance (precipitation and evaporation [Bradley 1999]). According to previous research, small basins with no inlets or outlets, such as those used in this study, are the best choices for creating a paleoclimate reconstruction (Forbes 1987, Stuiver 1970) on an extra-local scale.

By convention, the relative proportions of ^{18}O and ^{16}O are reported by calculating $\delta^{18}O$ for a substance, using the formula:

$$\delta^{18}O = 1000(R_{sample}/R_{standard} - 1)$$

where $R = {}^{18}O/{}^{16}O$ and the $R_{standard}$ is VSMOW-CO$_2$, with a $\delta^{18}O = 0$. This value is reported in per mil (‰) as a proportion of one thousand, analogous to the reporting of proportions of a hundred as percent (%). The more positive the $\delta^{18}O$ value, the more enriched it is in ^{18}O; the more negative the $\delta^{18}O$ the more depleted it is in ^{18}O.

Stable oxygen isotope ratios of lake sediments are controlled by several factors, such as the composition of lake water and the fractionation of isotopes during the formation of carbonates. The composition of the water in a lake is primarily dependent on two factors: the $\delta^{18}O$ of precipitation entering the lake and the $\delta^{18}O$ of water vapor leaving the basin through evaporation. Warm conditions deliver ^{18}O-enriched precipitation to a lake (Dansgaard 1964), leading to increased $\delta^{18}O$ in both the lake water and the carbonates created there. Warm dry conditions also result in increased evaporation from a lake, further enriching lake water in ^{18}O (Craig et al. 1963, Forbes 1987).

The isotopic composition of carbonates produced in a lake is a function of more than just the water chemistry (reservoir composition). The temperature of the water at the time of crystallization (McConnaughey 1986) and the energetics of the crystallization itself (Grossman and Ku 1986, Taruntani et al. 1969) also influence $\delta^{18}O$. Again, warmer conditions lead to greater concentrations of ^{18}O, and the $\delta^{18}O$ of carbonates increases with rising temperatures (Epstein et al. 1953, Forbes 1987, McConnaughey 1986). The physics of crystallization also enriches carbonates (the solid phase) in ^{18}O relative to the composition of lake water (the liquid phase [McConnaughey 1986]). This last factor is easy to control for because, all other things being equal, the disequilibrium between the solid and liquid phases is constant.

Previous Isotope Results from the Study Area

Forbes' research on Duley and Goose Lakes (Forbes 1987) confirmed that oxygen isotope records from lakes in the study area actually did track changes in climate. For example, the record from Duley Lake (Figure 3.8) showed an increase in $\delta^{18}O$ from about 11,000 to 9,500 years ago, consistent with the temperature rise that marked the transition from the Pleistocene to the Holocene. Likewise, he found that the composition

Goose Lake (Figure 3.7) was controlled by climate, with a sequence of increased $\delta^{18}O$ values corresponding to the known mid-Holocene period of aridity documented from other records (Forbes 1987).

Forbes has cautioned that changes in $\delta^{18}O$ from lakes in the study area could not be directly converted to a ratio-scale temperature record, even though work done on modern samples shows that an increase of $1^{\circ}C$ usually results in a 0.7 to 1.3‰ increase in $\delta^{18}O$ values of precipitation into some basins (Dansgaard 1964, Stuiver 1970). Instead, Forbes (1987) suggested that oxygen isotope records from small lakes in eastern Washington be considered more indicative of relative aridity. He reasoned that if temperature were the only factor determining the $\delta^{18}O$ of lake water, then lake water and precipitation recharging the lakes would be equal. He found, instead, that lake water was 10 per mil enriched in ^{18}O with respect to local precipitation, which could only happen if evaporation was preferentially removing the lighter isotope. Forbes (1987) calculated that it would take an annual loss of roughly 30% from the lake's volume to account for the observed enrichment; independent observations confirmed that a third of the water in Goose Lake was lost each year to evaporation (Forbes 1987, Washington State University 1975 in Forbes 1987). In lakes in the study area, Forbes showed that more positive, enriched values could be interpreted as warm, dry conditions, and negative values would be characteristic of cool, moist conditions. This conclusion has been reached elsewhere, such as on the Tibetan Plateau, where Wu and others (2007) documented changes in water composition that tracked climate, but pointed out that there was no linear relationship between $\delta^{18}O$ and air temperature. Like Forbes (1987), Wu and others (2007) suggested that $\delta^{18}O$ be used to reconstruct periods of relatively moist/cool conditions and warm/dry conditions. Consequently, no attempt will be made to convert $\delta^{18}O$ values directly to paleotemperature ($^{\circ}C$) values in the current study.

Isotope Records from Hidden and Rinker's Lakes

For this analysis, isotope samples were taken from the Hidden and Rinker's Lake cores. (Locations for these lakes are given in Figures 3.5, 4.1, and 4.2). Eleven marl samples of between 0.2 and 1.7 g (dry weight) were taken from each lake and submitted to Mountain Mass Spectrometry (Evergreen Colorado). A list of the twenty-two samples submitted and the resulting stable isotope measurements are reported in Table 5.1. Dates for each sample, in years B.P., were interpolated based on each sample's stratigraphic depth relative to the mean calibrated radiocarbon dates and sedimentation rates from Tables 4.2 and 4.3.

Individual samples were taken from a narrow range of stratigraphic depths, in order to minimize the span of time represented by any one sample. Each of the samples submitted spanned 5 mm, or approximately (11-13 years) of sedimentation, in the core. Besides sedimentation rate, the turnover rate of the lakes also had to be considered in

sampling. If these were large lakes, the water in the basin itself and its isotopic composition could be the product of 100 years or more of precipitation and evaporation, and thus would block any attempt to achieve a decadal-scale resolution. However, the turnover, or residence time, of the water in these shallow perennial basins is only one to three years (Forbes 1987) resulting in isotopic water composition that is representative, at any one point in time, of at most three years' worth of rain and evaporation. Each sample represents, at worst, a palimpsest of 16 years of climatic history. This satisfies the requirement that each data point span 50 years or less, so that it can be meaningfully compared with the already established paleodemographic record.

Results from the samples taken in the manner described are plotted in Figure 5.1. These data span the past 1,458 years, and results from both Hidden and Rinker's Lakes are displayed together. This plot shows that both records display short-term, centennial-scale, warming and drying episodes over the past thousand years. The high frequency of this variability shows that a decadal-scale signal is reflected in the data; any loss of resolution would have produced a smoother record.

Visual inspection of this graph also shows that, for any given point in time, the two lakes have different isotopic compositions, with Hidden Lake consistently more enriched in ^{18}O than Rinker's Lake. This is probably due to a difference in how the carbonates in the two lakes have mineralized. Carbonate, $CaCO_3$, can form different solid phases (such as dolomite, calcite, and aragonite). The energetics involved in the crystal formation of each phase differs, such that each phase has its own fractionation signature (Forbes 1987, Grossman and Ku 1986, Taruntani et al. 1969). Aragonite, for example, is isotopically heavy, incorporating proportionately more ^{18}O in its crystalline structure than calcite under equivalent conditions of water composition and temperature (Forbes 1987, Grossman and Ku 1986). Composition of aragonite itself can vary under constant conditions if different pathways are used to fix the oxygen. For example, photosynthetic algae and gastropods living in the same lake secrete biogenic aragonite with different $\delta^{18}O$ signatures; carbonates precipitated by algae tend to be more depleted in ^{18}O (have more negative $\delta^{18}O$ values) than carbonates in gastropod shells (Forbes 1987, Fritz and Poplawski 1974, Stuiver 1970)

Deposition of different carbonate phases in the two lakes is a reasonable explanation of the higher, relatively depleted $\delta^{18}O$ values seen at Hidden Lake (Figure 5.1). This is in agreement with the sediment descriptions from Hidden and Rinker's Lakes. Hidden Lake sediments had gastropod shells in the algal carbonates and lake marl, whereas Rinker's Lake did not contain visible gastropod shells, just algal carbonates and lake marl (Figures 4.4, 4.5). The lake with the gastropods (Hidden) should have higher $\delta^{18}O$ values, and the lake lacking gastropods (Rinker's) should have more negative $\delta^{18}O$ values; this is exactly what is seen in the results.

Table 5.1. Oxygen Isotope Sample Description and Results

Lake	Depth (cm b.i.)	Age B.P.*	Dry Weight (g)	$\delta^{18}O$ Reported by Lab (‰)
Hidden	0	0	0.2	-5.00
Hidden	5	105	1.0	-4.70
Hidden	20	210	0.5	-4.69
Hidden	15	314	0.8	-4.92
Hidden	20	419	0.6	-4.59
Hidden	25	524	0.7	-4.65
Hidden	30	629	0.7	-4.96
Hidden	35	734	0.8	-4.52
Hidden	40	839	0.5	-3.90
Hidden	45	943	0.6	-4.04
Hidden	50	1048	1.7	-3.94
Rinker's	0	0	0.4	-9.87
Rinker's	5	146	0.6	-8.47
Rinker's	20	292	0.5	-6.46
Rinker's	15	437	0.5	-6.71
Rinker's	20	583	0.7	-8.61
Rinker's	25	729	0.9	-8.24
Rinker's	30	875	1.1	-6.79
Rinker's	35	1020	1.0	-6.75
Rinker's	40	1166	1.2	-7.47
Rinker's	45	1311	1.6	-7.06
Rinker's	50	1458	1.1	-7.13

*based on interpolation from calibrated radiocarbon date

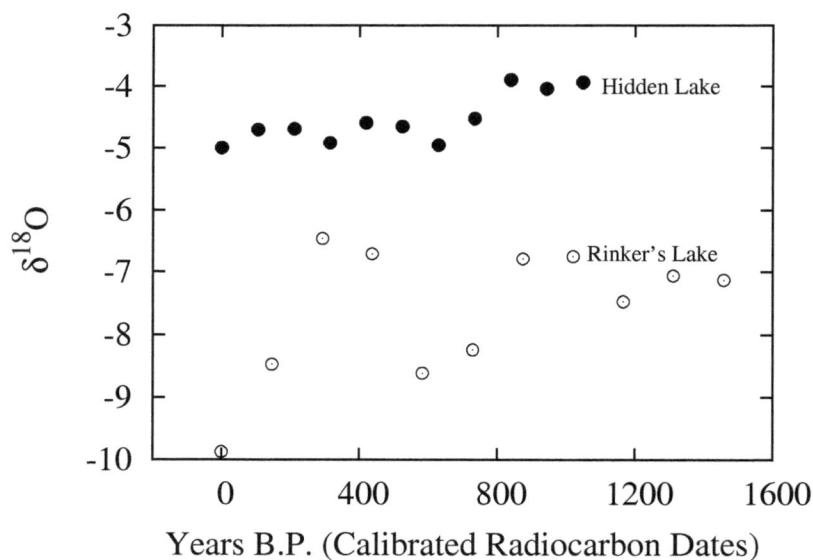

Figure 5.1. Oxygen Isotope $\delta^{18}O$ Results for Hidden and Rinker's Lakes

Forbes (1987), likewise, observed that Goose and Duley Lakes had different basal $\delta^{18}O$ values, but that the two lakes still reacted in kind to climatic fluctuations. Again, only one lake (Goose) contained gastropod remains, and this was the lake with more positive $\delta^{18}O$ baseline values (Forbes 1987). Based on Forbes' findings, Hidden and Rinker's Lakes can be expected to behave in a similar fashion; although starting with different isotope enrichment values, changes in $\delta^{18}O$ values both lakes can be expected to parallel one another. Visual inspection of the results from the two lakes (Figure 5.1) demonstrate that Hidden and Rinker's Lakes do indeed co-vary with one another, synchronously tracking changes in climate in the study area.

Combining Results into a Composite Isotope Index

Given that Hidden and Rinker's Lakes should be recording the same climatic fluctuations, it would be advantageous to combine the two records in some meaningful way into a composite index that reflects area trends over time. Although the samples are associated with different radiocarbon ages, linear interpolation can be used to estimate the values of the intervening points so that $\delta^{18}O$ can be estimated for coeval points from the two lakes. Interpolation between stable oxygen isotope samples can also be used to create a local climate record that is comparable to the existing paleodemographic record by calculating $\delta^{18}O$ values for each 50 cal-year interval from A.D. 950 to 1900. The resulting values are presented in Columns B and C of Table 5.2.

Another barrier to combining the two records is that the basal composition of the two lakes differed by several per mil. As seen in Figure 5.1 and Table 5.2, oxygen isotope values at Hidden Lake averaged -4.47‰, while those at Rinker's Lake averaged -7.53‰. Isotope values were mathematically transformed in order to compensate for this difference in isotope enrichment between the two lakes and allow information from both lake records to be combined into a single isotopic composition index for the study area. Observations from each lake were taken separately and re-expressed in terms of their deviations from the mean isotopic composition of the lake, and these values were standardized by dividing them by the standard deviation of $\delta^{18}O$ values from each lake. Use of these standardized deviations (or "Z-scores") for each observation removed the systematic bias (the 3.06‰ average difference between the lakes), and made it possible to meaningfully combine the isotopically enriched record from Hidden Lake with the relatively isotope-poor record from Rinker's Lake. Since all the original values were negative, the new composite data were multiplied by a factor of -1 to preserve the sign of the original observations. The composite index computed in this manner is given in Table 5.2 (Column F), and is plotted in Figures 5.2 (against calibrated B.P. dates) and 5.3 (fit with a spline and plotted against calendrical dates).

The composite isotope index from Hidden and Rinker's Lakes (Figure 5.2), suggests that climate conditions in the study area a thousand years ago were a little drier than today. The records indicate that, starting about a thousand years ago, there was a trend towards moister conditions which attained a local maximum at about 600 B.P., and then began to swing towards more xeric conditions. Based on the isotope data, this drying trend reached its local maximum at around 400 B.P., and was then followed by a final swing to more mesic conditions that continues until the present.

Relating this record to other regions and global climate, there is evidence for a local expression and recording of two short-term climatic events identified elsewhere as the "Medieval Warm Period" (MWP) and the "Little Ice Age" (LIA). The trend towards more mesic conditions, as seen in the Hidden and Rinker's lake cores, begins about 1,000 B.P. and ends around 600 B.P. (A.D. 950 to A.D. 1350). The timing of this event is similar to that of the "Medieval Warm Period," which was globally characterized as a slight rise in annual temperature that occurred between A.D. 900 and A.D. 1350 (Crowley and Lowery 2000, Grove and Switsur 1994, Hughes and Diaz 1994, Meese et al. 1994). In this study area, a coeval event is observed, but is expressed locally as a period of relatively moist conditions, not higher temperatures.

The second major feature seen in the Hidden and Rinker's Lake cores is an arid episode that roughly corresponds in time to the climatic fluctuation called the "Little Ice Age" (LIA). Previous work on the Plateau had shown that the LIA did not affect this interior region in the typical manner seen elsewhere (Chatters 1998). As it was globally expressed, the LIA was a period of lowered mean temperatures between the sixteenth and nineteenth, culminating in maximally cold conditions between A.D. 1570 and 1730 (Bradley and Jones 1993). In the records provided by Hidden and Rinker's Lake, there is a small climatic event that began about 600 B.P. (A.D. 1350), reached a maximum at 400 years ago (A.D. 1550) and ended in the last hundred years. This is similar in timing to the stereotypical characterization of the Little Ice Age, although in the study area, this climatic event seems to have a slightly later onset. The nature of this event is unusual in the study area as well; it is typified by more arid conditions rather than the cooler conditions seen both globally and in other regions. It seems that local manifestations of the LIA can run counter to global conditions, and lacustrine sediments from Devil's Lake on the Great Plains, likewise, record arid conditions during this period (Fritz et al. 1994).

Based on the $\delta^{18}O$ records from Hidden and Rinker's Lakes, climate conditions in the study area should be most favorable for the outbreak of natural fire when conditions are most xeric, and fuels are dry— conditions seen at 1,000 and 400 B.P. These times should also exhibit a relatively greater abundance of

Table 5.2. Interpolated Oxygen Isotope Values for each 50-cal Year Study Interval

Column A Date cal A.D.	Column B Hidden Lake Interpolated $\delta^{18}O$	Column C Rinker's Lake Interpolated $\delta^{18}O$	Column D Z-score for Hidden Lake Isotope Values	Column E Z-score for Rinker's Lake Isotope Values	Column F Composite Isotope Index*
1900	-4.70	-8.98	-0.58	-1.88	-1.23
1850	-4.70	-8.45	-0.58	-1.20	-0.89
1800	-4.70	-7.92	-0.58	-0.51	-0.55
1750	-4.80	-7.28	-0.84	0.33	-0.25
1700	-4.90	-6.63	-1.09	1.17	0.04
1650	-4.75	-6.62	-0.71	1.19	0.24
1600	-4.60	-6.60	-0.33	1.21	0.44
1550	-4.65	-6.84	-0.46	0.89	0.22
1500	-4.70	-7.08	-0.58	0.58	0.00
1450	-4.85	-7.69	-0.96	-0.20	-0.58
1400	-5.00	-8.29	-1.34	-0.99	-1.16
1350	-4.75	-8.37	-0.71	-1.08	-0.90
1300	-4.50	-8.44	-0.08	-1.18	-0.63
1250	-4.20	-8.26	0.67	-0.95	-0.14
1200	-3.90	-8.08	1.43	-0.71	0.36
1150	-3.95	-7.62	1.30	-0.11	0.59
1100	-4.00	-7.15	1.18	0.49	0.83
1050	-3.95	-6.96	1.30	0.74	1.02
1000	-3.90	-6.77	1.43	0.99	1.21
950	-3.85	-6.58	1.55	1.23	1.39
	$\bar{x} = -4.47$	$\bar{x} = -7.53$			
	$s_x = 0.40$	$s_x = 0.77$			

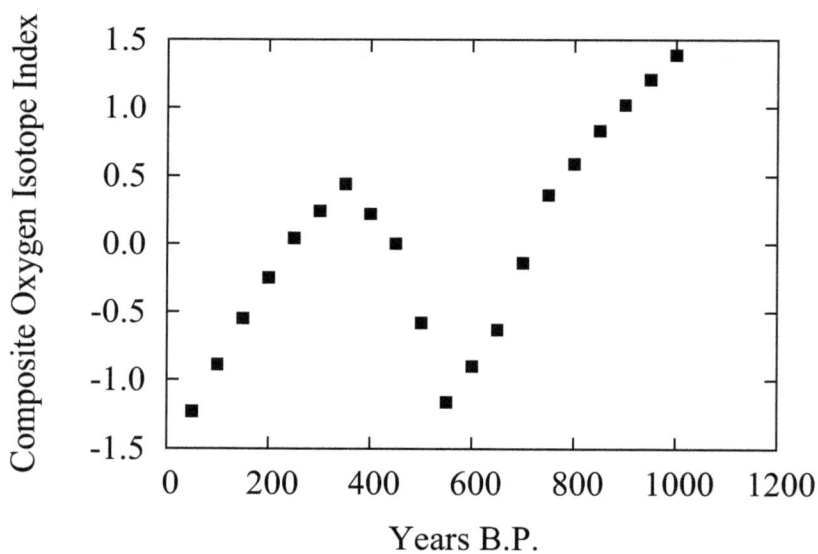

Figure 5.2. Composite Oxygen Isotope Index *vs.* Radiocarbon Years Before Present

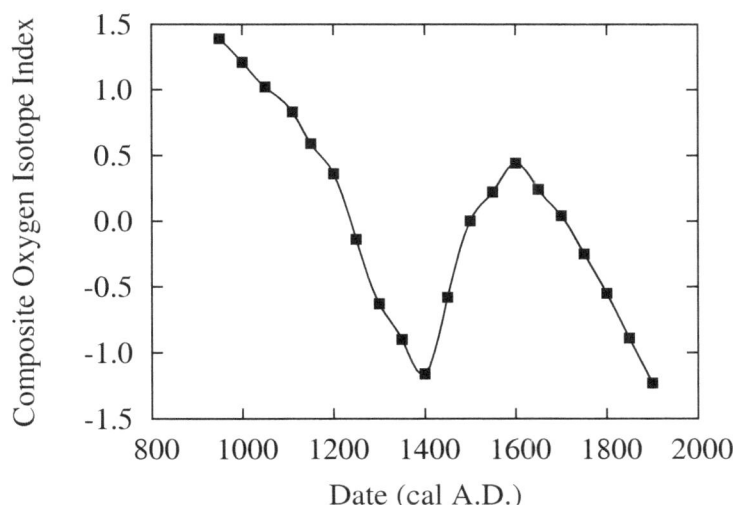

Figure 5.3. Composite Oxygen Isotope Index *vs*. Calendar Years A.D.

Table 5.3. Composite Oxygen Isotope Index Autocorrelations
(Correlations between the Composite Isotope Index and Future Values of Itself)

Significant Autocorrelations		
# of Lags	Correlation (Pearson's *r*)	Significance (*p*)
1 lag (50 years)	+0.910	< 0.0001
2 lags (100 years)	+0.658	< 0.003

Insignificant Autocorrelations		
# of Lags	Correlation (Pearson's *r*)	Significance (*p*)
3 lags (150 years)	+0.283	< 0.271
4 lags (200 years)	-0.132	< 0.626
5 lags (250 years)	-0.439	< 0.101
6 lags (300 years)	-0.592	< 0.026
7 lags (350 years)	-0.464	< 0.111
8 lags (400 years)	-0.159	< 0.623
9 lags (450 years)	+0.243	< 0.472
10 lags (500 years)	+0.718	< 0.019

taxa, such as ponderosa pine, that thrive in dry conditions with frequent fires. That is, if climate were the sole determinant of local conditions, $\delta^{18}O$ values should be positively correlated with charcoal influx and pine pollen values. Conversely, times of more negative $\delta^{18}O$ (e.g., 600 B.P.) which signal more available moisture, should favor more mesic vegetation and should support fewer natural fires. Trends opposite to this ($\delta^{18}O$ negatively correlated with charcoal influx) would be more likely if humans were determining the nature of the local fire regime.

Historical Constraints on Climate

How much is climate historically constrained in the study area, and on what time scales? To what extent is the oxygen isotope record conditioned by past values of itself? One way to examine the extent to which past climatic conditions influence future climatic conditions is to introduce lags into a correlation model. For a lag of 50 years, for instance, the isotope value from A.D. 1900 can be compared to A.D. 1850, the value from A.D. 1850 can be compared to A.D. 1800, and so on. In like manner, lags of 100, 150, 200 or even 500 years can also be introduced and correlations run between observations and the values that preceded them. Results of such an analysis are presented in Table 5.3.

As Table 5.3 shows, the oxygen isotope record is significantly and positively correlated with future values of itself when one (50-year) or two (100-year) lags are entered into the model. For longer time periods (150 to 300 years), there is no statistically significant relationship. As one might logically expect, climate is not truly a random variable but is historically constrained. However, the proxy record indicates that this effect only operates simply and clearly on a relatively short term basis, up to about a hundred years.

Chapter 6. Pollen Record (Vegetation Proxy) from Five Lakes

Geographical Context

Of all the basins used in this study, Five Lakes had the most organic sediments and was the most likely to produce a good pollen record. Five Lakes is located in Okanogan County, 9.5 km north of the Grand Coulee Dam and 5.5 km southeast of the modern town of Nespelem (Figure 3.5). The basin cored was the largest in the cluster of basins that make up Five Lakes, and a detailed map of the basins and coring location is given in Figure 4.3. Because the basin is closed and small, the pollen it receives is expected to come predominantly from vegetation sources within about a 10 km radius of the lake.

Figure 3.5 shows the proximity of both modern and prehistoric population centers to Five Lakes. A group of known prehistoric winter villages that cluster along a bend in the Columbia River lie 6 km southwest of the coring site—well within the pollen source area for the lake. Pollen being deposited in Five Lakes should literally be derived from the vegetation with which people have interacted for over a millennium.

A further advantage of using Five Lakes derives from its location at an elevation of 750 m asl. This places the lake near the modern elevational boundary between the pine woodlands and the open sagebrush-grass shrub steppe. Since the lake is located at the ecotone between major vegetation zones and is expected to provide an extralocal pollen signal, it should track subtle changes in the dominance and elevational distribution of woodlands versus steppe over time.

Five Lakes Pollen Dataset

The pollen counts from Five Lakes are recorded for reference purposes in Appendix F. These raw data were converted into percent abundances, which are also in Appendix F. Resulting relative abundances of major and minor taxa are plotted in Figure 6.1. The record is dominated by pine (*Pinus*), with lesser contributions from larch (*Larix*), Douglas fir (*Pseudotsuga*), hemlock (*Tsuga*), maple (*Acer*), oak (*Quercus*), birch (*Betula*), alder (*Alnus*), spruce (*Picea*), fir (*Abies*), members of the goosefoot family (Chenopodiaceae), sagebrush (*Artemesia*), other composites (Asteraceae), grasses (Poaceae family), and other rare herbaceous types (e.g., the Apiaceae and Polygonaceae). Of these, only a few taxa ever contribute significantly to the pollen percentage. If insignificant (rare) taxa are defined as those that never contribute over 5% to any pollen assemblage, then the significant major and minor taxa include only members of the following genera and families: *Pinus, Picea, Artemesia, Betula, Alnus,* Poaceae, Chenopodiaceae, and Asteraceae (including *Artemesia, Ambrosia, Aster,* and Liguliflorae). *Pinus* is by far the most abundant taxon, and the only type that consistently accounts for over 30% of the assemblages at any given depth.

Table 6.1 lists the taxa from Five Lakes, and divides them into categories of major, minor, or rare (insignificant) taxa. These designations are important from a vegetation (community) standpoint as well as from a statistical standpoint. Rare taxa occur so infrequently in the 500-grain pollen counts used in this study that the error of estimating their abundance is too great to allow their measured percent contribution to the pollen spectra to be statistically reliable. Statistical tests presented in this and future chapters will therefore ignore the role of rare taxa. *Sporormiella* and other spores were also identified during pollen counting. These, too, were rare and dropped from further analysis.

Interpretation of Pollen Results

The most striking aspect of the pollen percentage diagram (shown in Figure 6.1) is that the record of the past 1,525 years has been dominated by *Pinus* throughout. Looking at the record as a whole, several trends are evident. Figure 6.1 shows that *Pinus* decreases in abundance over time, while the relative contributions of *Betula*, Chenopodiaceae, Poaceae, *Artemesia*, and other Asteraceae simultaneously increase. Statistics confirm these impressions, and correlations between the abundance of each taxon and depth are presented in Tables 6.2 (for significant correlations) and 6.3 (for insignificant correlations). These data demonstrate that the taxa listed above do show significant directional changes in percent abundance; *Betula*, Chenopodiaceae, Poacea, *Artemesia*, and some other Asteraceae (*Aster* and *Ambrosia*) are all significantly negatively related with depth, while *Pinus* is positively correlated with depth.

Although most of the taxa show a directional change throughout the core, even this overall trend can be broken down further into two separate zones within the pollen record. Visual inspection of the pollen diagram in Figure 6.1 indicates that there is a lower zone extending from about 50 to 145 cm b.i. (593 to 1,525 B.P.), characterized by high and constant domination by *Pinus*. In contrast, there is an upper zone characterized by a lower and more variable abundance of *Pinus* starting around 50 cm b.i. and extending to the top of the core (0 cm b.i.).

Pollen data were submitted to Principal Components Analysis (PCA) using SPSS v.10, and results are listed in Tables 6.4 and 6.5. Figure 6.2 presents samples plotted as a function of the two main multivariate axes, showing the placement of the successive samples relative to each other in multidimensional space. The graph suggests that there are two main spaces being occupied by the samples. These two multidimensional spaces are shown as shaded areas in Figure 6.3, an annotated version of the same plot and data provided in Figure 6.2.

Table 6.1. List of Taxa from Five Lakes Pollen Record

Characterization of Taxa from Five Lakes Pollen Record
Dominant Taxon (most abundant—over 30% of any sample)
Pinus
Minor Taxa (significant contributors under 20% of any sample)
Alnus
Picea
Betula
Poaceae
Chenopodiaceae
Asteraceae:
Artemesia
Ambrosia
Aster
Ligulaflorae
Rare Taxa* (never contribute 5% or more to any sample)
Apiaceae
Larix/Pseudotsuga
Abies
Quercus
Acer
Tsuga
Polygonaceae

*rare taxa have such low percent abundances that their contributions cannot be estimated with statistical confidence. These taxa should be ignored in statistical tests.

Table 6.2. Significant Correlations between Taxa and Sample Depth from Five Lakes

Positive Correlations

Taxon/Taxonomic Group	Correlation (Pearson's r)	Significance (p)
Pinus	+0.730	< 0.0001
Picea	+0.585	< 0.001
AP	+0.764	< 0.0001

Negative Correlations

Taxon/Taxonomic Group	Correlation (Pearson's r)	Significance (p)
Betula	-0.658	< 0.0001
Alnus	-0.523	< 0.003
Chenopodiaceae	-0.588	< 0.001
Poaceae	-0.484	< 0.007
Artemesia	-0.711	< 0.0001
Ambrosia	-0.726	< 0.0001
Aster	-0.696	< 0.0001
NAP	-0.764	< 0.0001

Table 6.3. Insignificant Correlations between Taxa and Sample Depth from Five Lakes

Negative Correlations*

Taxon/Taxonomic Group	Correlation (Pearson's r)	Significance (p)
Liguliflorae	-0.317	< 0.088

*there are no insignificant positive correlations

Table 6.4. PCA Scores for Pollen Samples

Sample Depth cm b.i.	First Component Score	Second Component Score
0	0.446	0.247
5	0.173	0.003
10	0.000	0.000
15	0.394	0.162
20	0.216	0.126
25	0.247	0.054
30	0.221	0.396
35	0.310	0.276
40	0.410	0.276
45	0.539	0.185
50	0.615	0.243
55	0.542	0.177
60	0.770	0.301
65	0.694	0.266
70	0.666	0.224
75	0.787	0.290
80	0.743	0.283
85	0.761	0.297
90	0.689	0.279
95	0.654	0.281
100	0.654	0.246
105	0.691	0.224
110	0.624	0.271
115	0.612	0.191
120	0.654	0.231
125	0.593	0.239
130	0.703	0.259
135	0.707	0.255
140	0.741	0.261
145	0.779	0.303

Table 6.5. PCA Scores for Taxa

Taxon	First Component Score	Second Component Score
Pinus	0.886	0.344
Betula	-1.179	-1.962
Alnus	-0.129	-0.949
Picea	0.869	0.464
Chenopodiaceae	-0.807	1.060
Poaceae	-0.180	0.253
Artemesia	-0.803	-0.805
Ambrosia	-1.157	1.464
Aster	-1.243	1.507
Liguliflorae	-0.535	-1.894

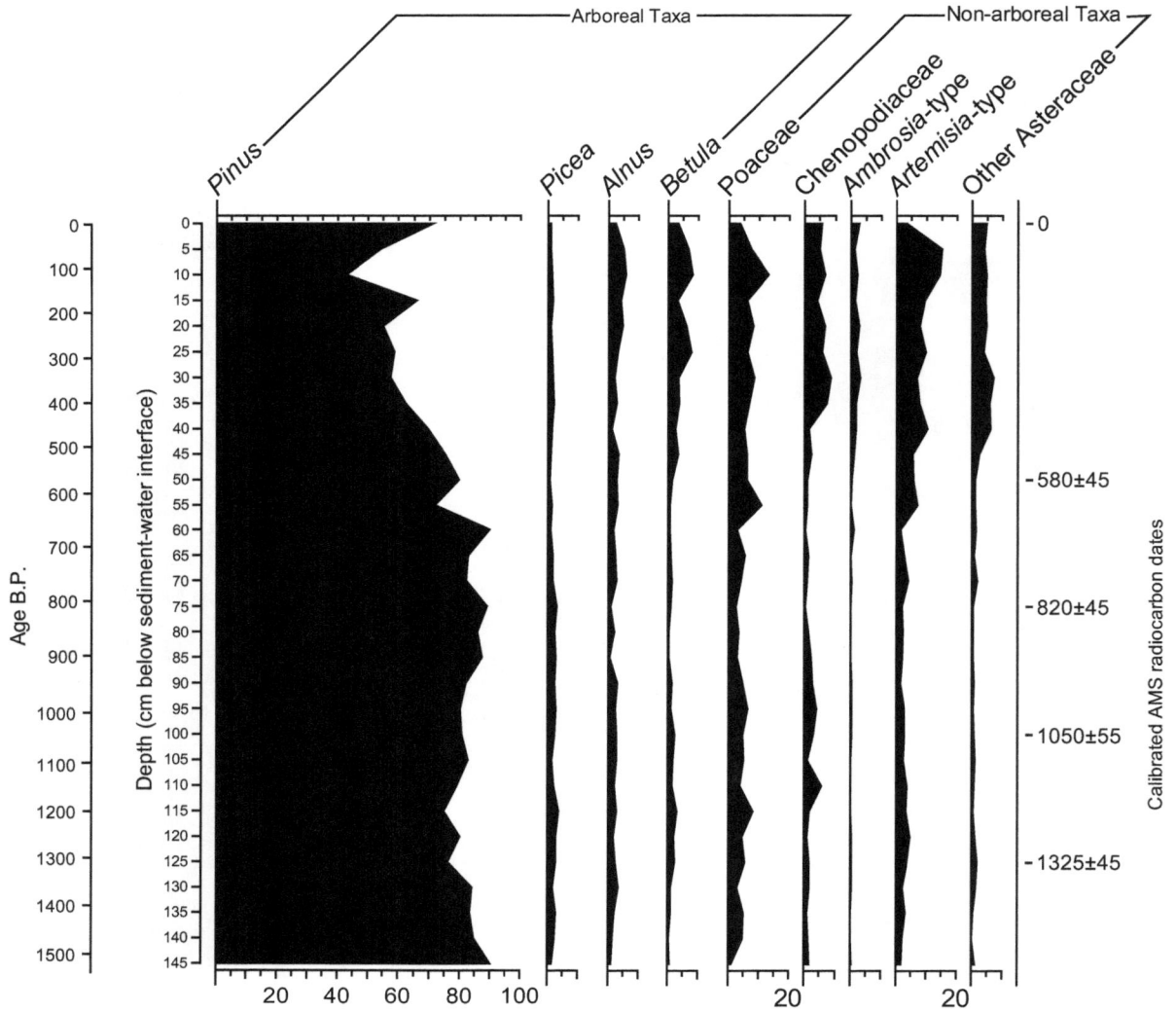

Figure 6.1. Pollen Diagram for Five Lakes, OK Co., WA
(Other Asteraceae = *Aster*-type and Liguliflorae)

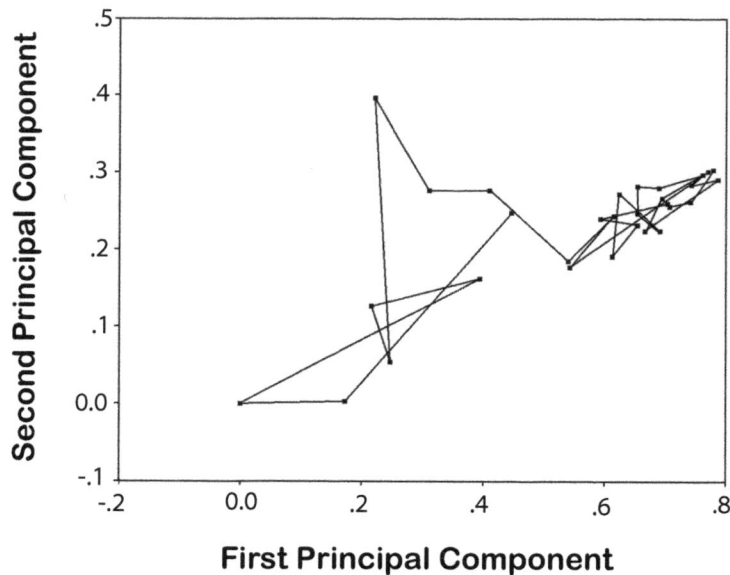

Figure 6.2. First and Second Principal Components Derived from Pollen Spectra

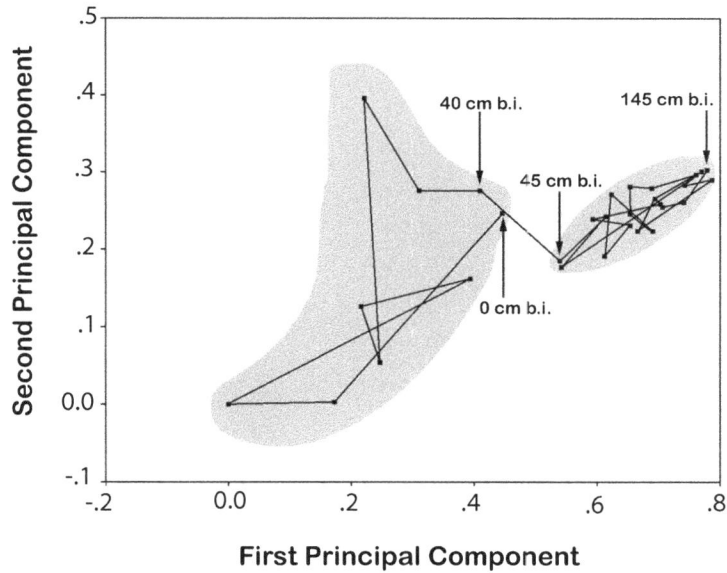

Figure 6.3. Annotated First and Second Principal Components

Figure 6.3 shows that the series of samples in multidimensional space can be easily divided into two separate sets of contiguous samples. One coherent set of points is made up of samples from the upper zone of 0-40 cm b.i., as seen grouped together in the shaded area in the left side of Figure 6.3. A second coherent set of points is made up of the more closely spaced set samples that are all drawn from 45 to 145 cm b.i. (as seen at the right in Figure 6.3).

The two zones differ in two main ways—in their abundance of arboreal (mostly pine) pollen and in the stability of the pollen types within the zones. The lower zone from 145 to 45 cm b.i. corresponds to a time period from roughly 1,525 to 529 B.P. This zone contains a high and consistent domination of *Pinus* at an average of 80% throughout the zone. Non-arboreal taxa—shrubs and herbaceous plants—account for a small percentage of the assemblage, averaging only 11% of the pollen sum for the whole time period. In comparison, the upper, more recent pollen zone shows a significantly lower and more variable amount of pine, with an average of only 60% of all pollen grains coming from the genus *Pinus* (see Figure 6.1).

This difference can be interpreted in several ways. This trend in pine and non-arboreal pollen types indicates that during the early period, pine was relatively more abundant than it is today. Pine was more common on the landscape and pine forests covered a greater areal extent and extended into lower elevations at that time than they do today. This also implies that herbaceous and shrubby taxa were relatively less common and less important during the early part of this record. Plots for various taxa in Figure 6.1 indicate this, and a graph of the relative abundance of all non-arboreal types with depth given in Figure 6.4 confirms this.

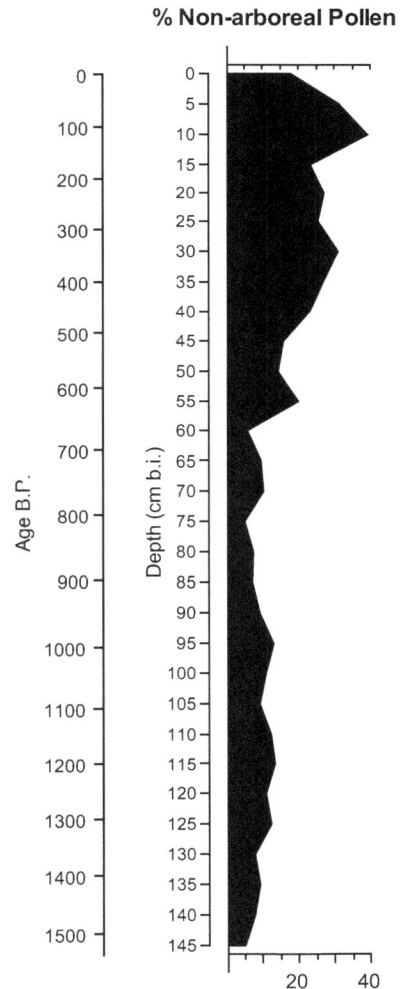

Figure 6.4. Percent of Non-arboreal Pollen with Stratigraphic Depth

49

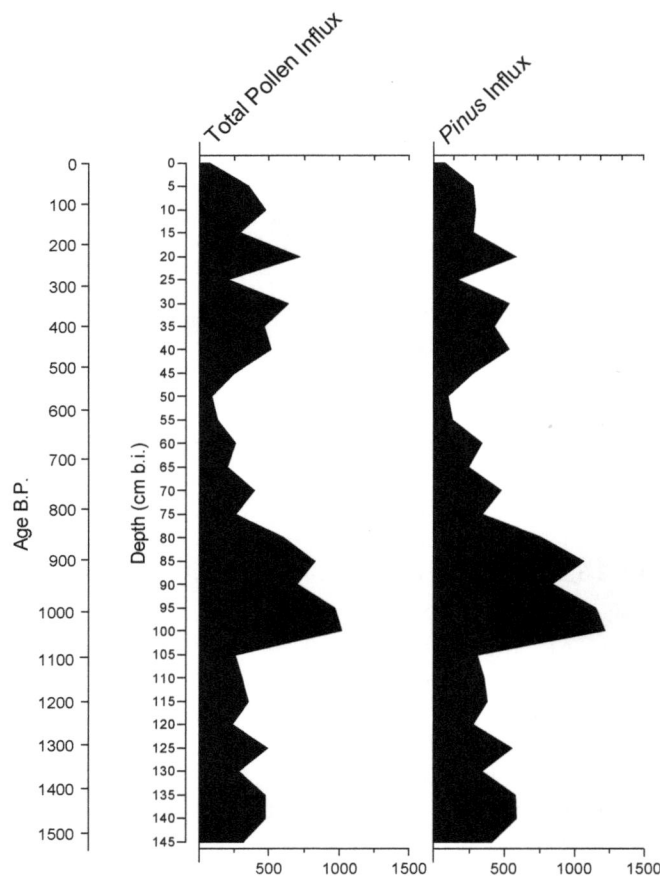

Figure 6.5. Absolute Influx* of for all Pollen Grains (left) and Pine Grains (right)
*Pollen Accumulation Rate (PAR) in grains/cm^2/yr

This is also confirmed by the pollen accumulation rates (PAR, measured in grains/cm^2/yr), which is an absolute measure of pollen deposition. *Pinus* pollen influx can be taken as a proxy measure for the number of trees of this genus present on the landscape. Figure 6.5 (based on data in Appendix F) shows the influx of pine pollen, indicating that there were more pine pollen grains being produced and deposited during the early part of the pollen record. Likewise, PAR for non-arboreal pollen (shown in Figure 6.6, from data in Appendix F) confirms that the lower relative abundance of non-arboreal taxa in the lower zone is a reflection of a lower absolute number of non-arboreal pollen grains being deposited, probably because there were fewer of these plants on the landscape.

Other than being more heavily dominated by pine, the lower zone also has a less variable taxonomic composition and representation—there are no smaller scale low-amplitude or low-frequency events that can be resolved. The upper zone shows a less stable vegetation, with higher variability within the zone. This was suggested by the plot of samples in multidimensional space, shown in Figures 6.2 and 6.3. These figures show that the points representing samples in the lower zone are

more closely clustered together than are the points for samples from the upper zone. Obviously, the samples from the lower zone are more coherent and more similar to each other than are the samples from the upper zone.

For greater clarity, the same information gleaned from principal components analysis can be transformed and presented in a simplified manner using detrended correspondence analysis (DCA) (Gauch 1982, Jacobson and Grimm 1986). DCA results are in Figure 6.7, with detrended scores from the first axis (which explains over 80% of all variability) plotted against sample depth. This shows that the scores change little between 145 and about 60 cm depth in the core. Starting at 60 cm b.i. and proceeding up the core, the values begin to rise and fall with a greater amplitude, indicating greater rates of change between samples in the 60 to 0 cm b.i. portion of the core.

An examination of the pollen zones and the changing abundances of pollen taxa over time suggests that certain sets of taxa co-vary with one another. A calculation of the correlation between taxa, as measured by Pearson's *r*, is given in Tables 6.6 and 6.7. Table 6.6 lists all statistically significant ($p \leq 0.01$) correlations between

NAP influx

Score / DCA Loading

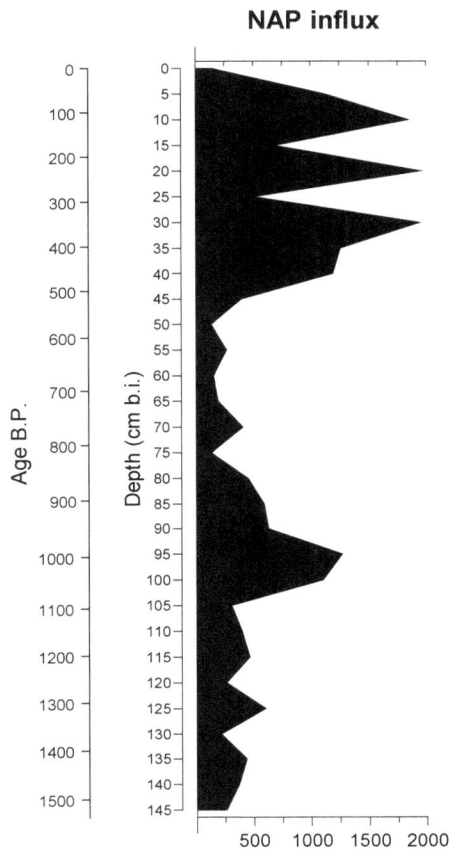

Figure 6.6. Absolute Influx* for Non-arboreal Pollen
*Pollen Accumulation Rate (PAR) in grains/cm²/yr

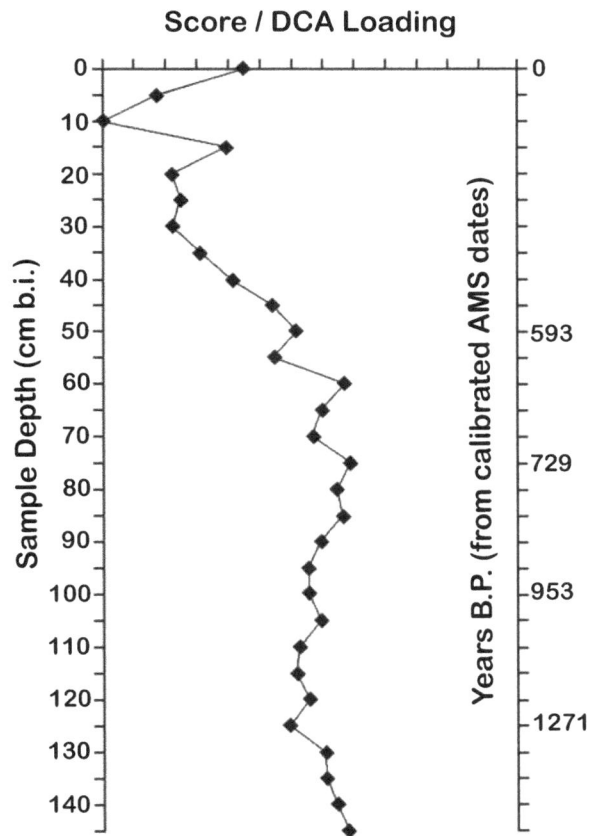

Figure 6.7. DCA* (Detrended Correspondence Analysis) Scores by Depth and Time
(*DCA values calculated using MVSP v3.1)

taxa, and this list reveals important taxonomic relationships and structure within the dataset.

As one might expect, the dominant taxon, *Pinus*, is negatively correlated with nearly all of the minor taxa (Tables 6.1 and 6.6), since a decrease in *Pinus* on the landscape opens the land for colonization by other plants. *Pinus* is significantly negatively correlated with many of the important minor taxa such as Chenopodiaceae, Poaceae, *Alnus*, *Betula*, and most of the members of the Asteraceae family (*Artemesia*, *Aster*, and *Ambrosia*). This confirms that pines expand at the expense of more seral herbaceous and shrubby species, and that decreases in pine woodlands encourage the expansion of more open "disturbance" vegetation.

Among the non-dominant taxa are several pairs of minor (but not rare) taxa in the pollen diagram that co-vary with each other. For example, *Alnus* and Poaceae are significantly positively correlated with each other, as are *Betula* and Chenopodiaceae, *Betula* and *Artemesia*, as well as Poaceae and *Artemesia*. The PCA results confirm this; the scores for all these taxa (*Alnus*, *Betula* Chenopodiaceae *Artemesia, Ambrosia,* and *Aster*) on the first principal component are very similar in both sign and magnitude, and this dimension explains more than 60% of the variability in all samples over time (Table 6.5). All of these taxa are considered to be indicative of

disturbance or land clearance when found in pollen diagrams in the Northwest. And, indeed, the bivariate correlations and PCA confirm that these taxa form a coherent suite of taxa. When any one of these pollen types increases, it signals a change in an underlying disturbance variable and signals changes in the amount of disturbed, open land.

Again, all of these indicators are high in value in the upper 50 cm of the core, indicating a period of relatively high disturbance and a recession of wooded areas in the past 593 years. This confirms the results of previous pollen studies that indicated a rise in disturbance taxa, such as Chenopodiaceae, in recent (historic) times. It is at odds, however, with the exact magnitude and timing of similar events seen in other pollen cores in the interior Northwest. For example, at Wildcat Lake, the historic rise in Chenopodiaceae was more substantial, with Chenopodiaceae accounting for a full 47% of the modern pollen spectrum (Davis et al. 1977). At Five Lakes, modern Chenopodiaceae never exceed 10% of the entire assemblage at any time over the past 1,525 years. At Wildcat Lake, disturbance species increase starting about 100 years ago (Davis et al. 1977); at Five Lakes the disturbance indicators rise much earlier, around 593 years ago.

Table 6.6. Significant Correlations between Taxa from Five Lakes

Positive Correlations

Taxonomic Pairs	Correlation (Pearson's *r*)	Significance (*p*)
Pinus-Picea	+0.352	< 0.0001
Betula-Alnus	+0.742	< 0.0001
Betula-Chenopodiaceae	+0.695	< 0.0001
Betula-Poaceae	+0.627	< 0.0001
Betula-Artemesia	+0.814	< 0.0001
Betula-Ambrosia	+0.664	< 0.0001
Betula-Aster	+0.588	< 0.001
Alnus-Chenopodiaceae	+0.475	< 0.0001
Alnus-Poaceae	+0.719	< 0.0001
Alnus-Artemesia	+0.670	< 0.0001
Chenopodiaceae-*Artemesia*	+0.532	< 0.002
Chenopodiaceae-*Ambrosia*	+0.761	< 0.0001
Chenopodiaceae-*Aster*	+0.626	< 0.0001
Poaceae-Artemesia	+0.689	< 0.0001
Artemesia-Ambrosia	+0.546	< 0.002
Artemesia-Aster	+0.768	< 0.0001

Negative Correlations

Taxonomic Pairs	Correlation (Pearson's *r*)	Significance (*p*)
Pinus-Betula	-0.908	< 0.0001
Pinus-Alnus	-0.746	< 0.0001
Pinus-Chenopodiaceae	-0.785	< 0.0001
Pinus-Poaceae	-0.793	< 0.0001
Pinus-Artemesia	-0.892	< 0.0001
Pinus-Ambrosia	-0.748	< 0.0001
Pinus-Aster	-0.787	< 0.0001
Betula-Picea	-0.318	< 0.0001
Alnus-Picea	-0.400	< 0.0001
Picea-Chenopodiaceae	-0.221	< 0.0001
Picea-Ambrosia	-0.462	< 0.010

Table 6.7. Insignificant Correlations between Taxa from Five Lakes

Positive Correlations

Taxonomic Pairs	Correlation (Pearson's *r*)	Significance (*p*)
Betula-Liguliflorae	+0.284	< 0.128
Alnus-Ambrosia	+0.382	< 0.037
Alnus-Aster	+0.356	< 0.053
Alnus-Liguliflorae	+0.244	< 0.193
Chenopodiaceae-Ligulaflorae	+0.216	< 0.252
Poaceae-*Ambrosia*	+0.400	< 0.029
Poaceae-*Aster*	+0.424	< 0.019
Poaceae-Ligulaflorae	+0.177	< 0.349
Artemesia-Ligulaflorae	+0.325	< 0.079
Ambrosia-Ligulaflorae	+0.215	< 0.255
Aster-Ligulaflorae	+0.236	< 0.210

Negative Correlations

Taxonomic Pairs	Correlation (Pearson's *r*)	Significance (*p*)
Pinus-Liguliflorae	-0.299	< 0.109
Picea-Poaceae	-0.173	< 0.362
Picea-Artemesia	-0.422	< 0.020
Picea-Aster	-0.430	< 0.018
Picea-Liguliflorae	-0.101	< 0.595

Overall, then, the pollen record at Five Lakes follows the trends expected from earlier pollen work in the region. As expected, the entire late prehistoric period is dominated by arboreal pollen, especially pine. This is no surprise, as Mack et al. (1976) characterized the whole of the late Holocene (past 4,000 years) in the Columbia Basin as a ponderosa pine-dominated period. Previous pollen records for the study area itself, from Rex Grange and Goose Lakes (Figures 3.6 and 2.5), demonstrated that pine abundances were high during the entire time spanned by the Five Lakes core.

Earlier in the Five Lakes sequence, pines are maximally abundant. As ponderosa pine is common in this area, and fire adapted, one might expect this earlier portion of the core to be a time of greater fire frequency. Also, as expected, the historic period shows an increase in the abundance and absolute influx of disturbance, steppe, and understory taxa such as *Artemesia*, Asteraceae, *Betula*, and Chenopodiaceae. This is traditionally viewed as being due to historic human disturbances (Davis et al. 1977), but could also be the result of natural disturbances, or environmental changes such as changes in fire frequency and/or available moisture. If human disturbance is implicated, the relatively high percentage of these indicators between 593 B.P. and 100 B.P. shows that Native Americans were perhaps influencing the local vegetation during the late prehistoric and Protohistoric to the same extent, and in the same way, as historic and Euro-American land use practices. Also, the sustained (but lower) presence of these seral taxa before 593 B.P. argues for palpable, although low magnitude, prehistoric anthropogenic influence predating 593 B.P. as well. Comparisons of the pollen data set with climate, fire, and population proxies promise to clarify these issues.

Chapter 7. Charcoal Record (Fire Proxy) from Five Lakes

Preliminary laboratory analyses identified Five Lakes as a good candidate for charcoal analysis. All fire proxy data were derived from Five Lakes sediments, and charcoal counts were done under the microscope at the same time as the pollen counts. The resulting raw charcoal counts, made for three different size classes, are recorded in Appendix G.

Figure 7.1 presents the absolute charcoal influx (CHAR) throughout the core, using two common forms of quantification used in the literature. On the left side of Figure 7.1, CHAR is displayed using the total number of particles deposited per square centimeter per year ($\#/cm^2/yr$ [e.g., Mehringer et al. 1977a, Millspaugh and Whitlock 1995, Whitlock and Millspaugh 1996]). Although this is the most popular manner of presenting influx data, charcoal influx is also often expressed in terms of the $mm^2/cm^2/yr$ deposited in a lake (Clark and Royall 1995a, 1995b). This second measure is plotted on the right, next to the plot of $\#/cm^2/yr$ in Figure 7.1 so that the results from these two techniques can be easily compared. Simple visual inspection of the plots in Figure 7.1 shows that both measures indicate that charcoal was deposited in greater amounts during the lower part of the core, and the charcoal influx plots exhibit many smaller

synchronous peaks (local maxima) through the rest of the core. Both popular methods of expressing charcoal influx ($\#/cm^2/yr$ and $mm^2/cm^2/yr$), as plotted in Figure 7.1, show nearly identical trends and features over time. Statistical analysis confirms that these methods are tracking the same trends as they are correlated with an $r = +0.943$ and $p < 0.0001$. This indicates that the charcoal results are robust—the same trends manifest themselves no matter how the charcoal is quantified. As $\#/cm^2/yr$ is the most popular means for expressing CHAR, it is used in the rest of the current analysis.

Figure 7.2 shows the total CHAR broken down into three different size classes. The smallest size class, 75-125 μm, accounts for most of the influx throughout the record, as expected by previous research (e.g., Clarke 1988a). The 125-250 micron class contributes a variable but significant amount, while the largest size class (250-500 μm) contributes little to any given sample. One trend evident from Figure 7.2 is that the representation of different particle size fractions changes over time, with larger diameter particles (125 to 500 μm-diameter) being deposited in significant numbers early on (low in the core), then declining and virtually disappearing around 500 B.P. (about 45 cm b.i.). This is of interest because

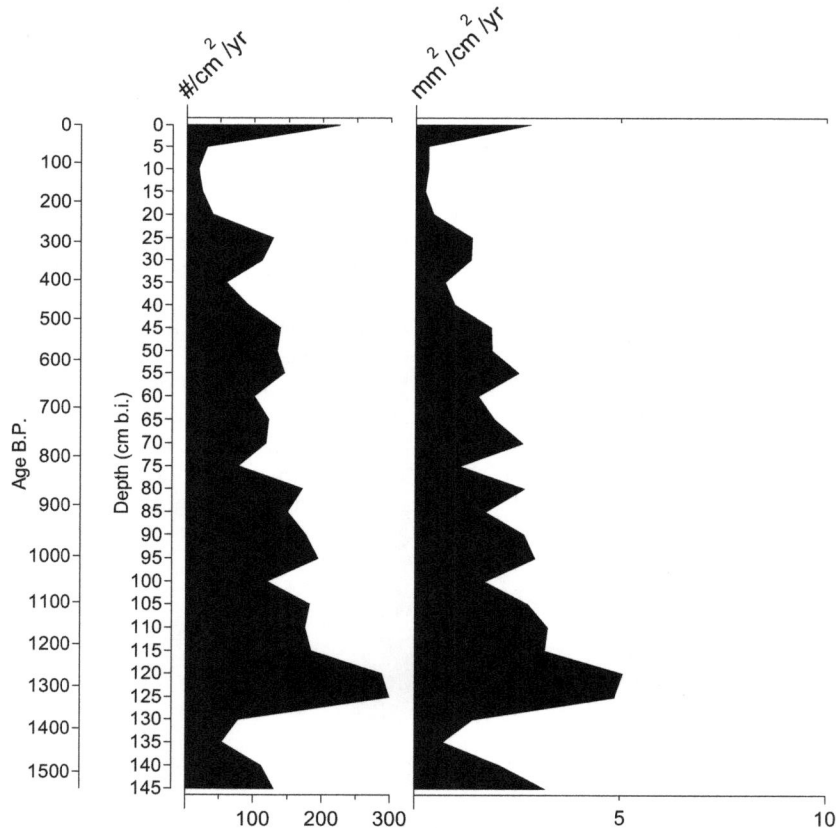

Figure 7.1. Total Charcoal Influx Presented Following Two Conventions at Five Lakes
($\#/cm^2/yr$ at left and $mm^2/cm^2/yr$ at right)

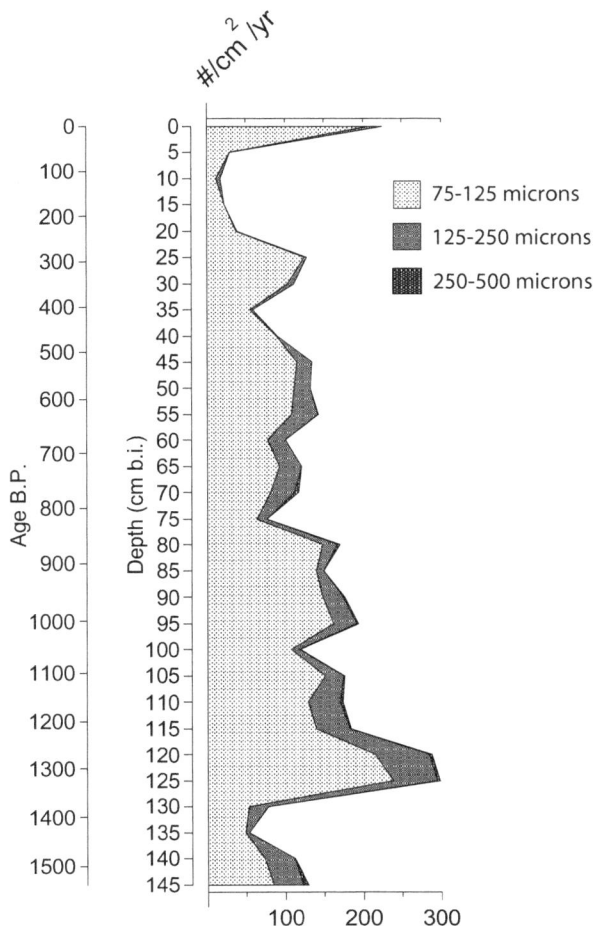

Figure 7.2. Charcoal Particle Influx (#/cm²/yr) for Five Lakes

charcoal particle size is a clue to the transport mechanism and distance traveled by each particle (Clark 1988a, Patterson et al. 1987). Trends seen in the plot suggest that earlier contributions to the charcoal record came from both regional and extralocal fires, while the more recent sediments received charred materials primarily from regional sources. Since this study poses questions on a local and extralocal scale, it is advantageous to use particle size as a way to focus on only those portions of the charcoal record that are extralocal, but not regional or global, in scale. Charcoal influx, when partitioned by particle size, is an excellent means for isolating signals reflecting different sources and spatial scales.

Extralocal and Local Charcoal

An extralocal signal is best characterized by the influx of large charcoal particles that could have traveled only a short distance (Clark 1988a, Clark and Hussey 1996, Clark and Royall 1995b, Morrison 1994, Patterson et al. 1987). Generally, researchers believe that charcoal from 1 to 125 µm in size can easily travel long distances in suspension high in the air column. Small charcoal (under 125 µm) is assumed to represent material primarily from regional, continental, or even global source areas and is often used by researchers analyzing large-scale temporal

and spatial processes (Carcaillet et al. 2001, Clark 1988b, Clark and Royall 1995b, Patterson et al. 1987, Whitlock and Larsen 2001). Since larger particles (those over 125 µm in diameter) only move over relatively short distances, charcoal records composed solely of large charcoal grains are assumed to reflect processes occurring on a smaller-than-regional-scale distances (Carcaillet et al. 2001, Clark 1988b, Clark and Royall 1995b, Gardner and Whitlock 2001, Whitlock and Larsen 2001).

Studies of charcoal as a proxy fire record, when checked against historical and fire scar records of vegetation fires, confirm the relationship between particle size and transport distance. Several studies have shown that small particles, under 125 µm in diameter, did not correlate with known fires in the extralocal area (Clark 1988a, Whitlock and Millspaugh 1996). Larger particles between 125 and 250 µm, however, provided records that corresponded with known fire events within the catchment of a lake (Clark 1988a; Clark and Royall 1995b; Whitlock and Millspaugh 1996). A separate examination of different size classes from Five Lakes, then, should tease apart fire proxies with different spatial-scale signals.

Large (125-500 µm) Particles

Figure 7.3 shows a graph of charcoal influx for all particles larger than 125 µm in diameter, assumed to be produced by fires that burned relatively close to the basin. Figure 7.3 shows a series of about eleven peaks of varying size over the 1,525-year record covered by the Five Lakes core. As peaks are commonly interpreted as fire events (e.g., Marlon et al. 2006), these features are labeled in Figure 7.4 (which is an annotated version of Figure 7.3). All local maxima on the graph are identified as fire events for the purposes of this research. Figures 7.3 and 7.4 indicate that there are eleven separate significant fire events evident in the Five Lakes charcoal record. These events, along with their associated depths and ages, are listed in Table 7.1.

One trend that is evident from the large-particle influx plot is that there is a systematic change in the charcoal influx over time. In Figure 7.3, the three most recent fire events (from 40 to 0 cm b.i.) are the smallest in the core. In contrast, both the background influx of large charcoal particles, and the local maxima, are greatest from 40 cm b.i. to the lowest point in the core. The sediments from Five Lakes, then, record a drastic decrease in large charcoal particle deposition into the lake starting around 500 B.P. and continuing to the present day.

Of greater interest than amplitude is peak frequency. The number of years that have elapsed between peaks is referred to in the fire literature as the return interval between fires (McBride 1983). This information, along with the change in return intervals, is listed in Table 7.1. As Table 7.1 shows, the return interval between fire events averages 148.4 years and ranges from 94 to 232 years. A plot of the return intervals associated with each fire event is given in graphic form in Figure 7.5. As both

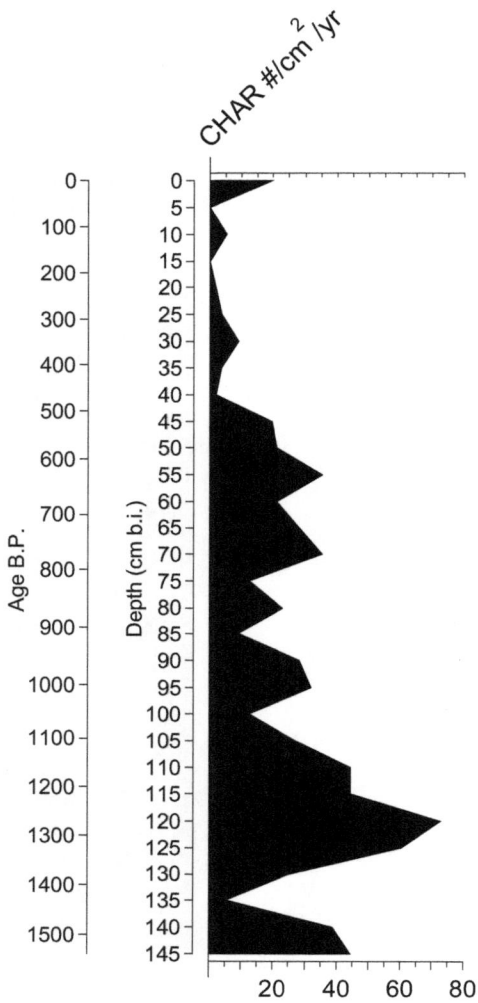

Figure 7.3. Influx (#/cm²/yr) of Large (>125 μm) Charcoal Particles at Five Lakes

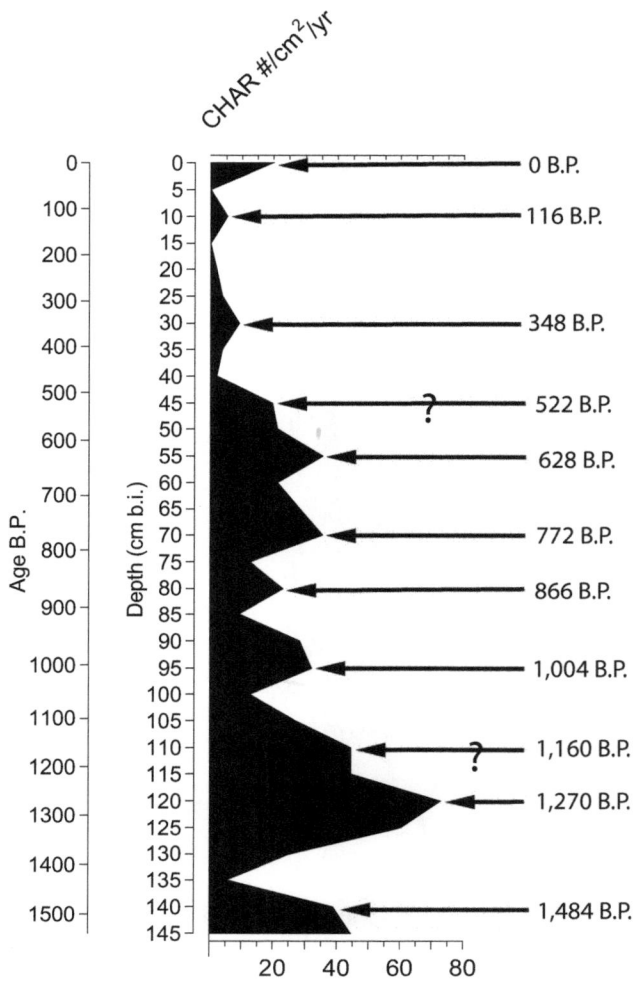

Figure 7.4. Annotated Influx (#/cm²/yr) of Large (>125 μm) Charcoal Particles for Five Lakes

Note: Question marks ("?") are used to denote peaks that are unclear, flat, or appear as shoulders on larger peaks.

Table 7.1. Fire Events from Large (125-500 μm) Charcoal

Sample Depth cm b.i.	Estimated Date of Event (B.P.)	Return Interval (yrs)	Change in Return Interval (yrs)
0	0	116	-116
10	116	232	58
30	348	174	68
45	522	106	-38
55	628	144	50
70	772	94	-44
80	866	138	-18
95	1,004	156	46
110	1,160	110	-104
120	1,270	214	
140	1,484	N.A.	
Average: 148.4			

Return Interval (in years)

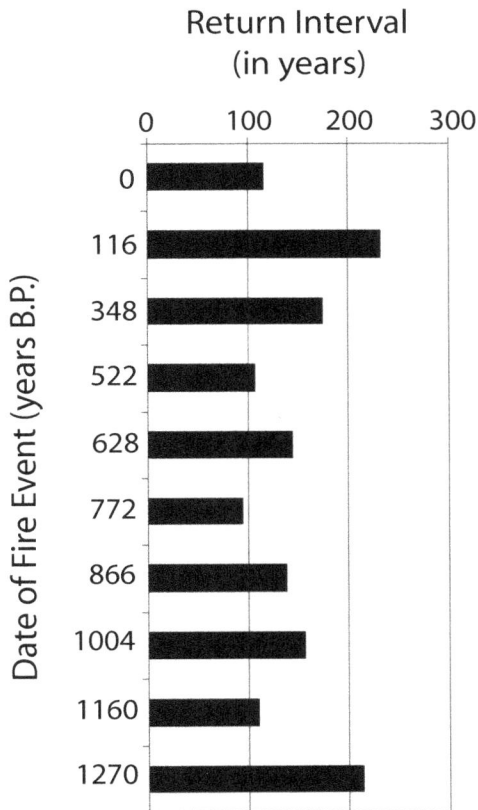

Figure 7.5. Fire Return Intervals at Five Lakes (Based on Peaks in 125- 250 μm Charcoal Influx Plot)

the plot and table show, there is no obvious pattern to the return intervals; the charcoal record at Five Lakes cannot be subdivided into periods of greater or lesser fire frequencies.

The influx of large particles can also be used to investigate if, and how, the outbreak of past fires influences future fires in an area. Table 7.2 shows the results of running correlations between past and future charcoal influx values with successively longer lags introduced into the model (in 50-year increments). Since vegetation fires consume the available fuel on the landscape, one might expect fire to be negatively correlated with itself over the short term. The opposite trend is seen in the charcoal dataset, however. Statistically significant autocorrelations do exist over the short term, but these are positive relationships. This could be due to the processes structuring the proxy itself, and not indicative of trends in the underlying nature of the area fires. For example, positive correlations between two consecutive charcoal samples could be due to the use of linear interpolation (which by its nature produces new interpolated points that are mathematically related to both neighboring samples). Positive correlations between successive charcoal samples are also likely to be due to the fact that all charcoal from a fire is not instantaneously deposited in a lake but arrives for many years after the event (Whitlock and Millspaugh 1996). Delayed transportation and reworking of charcoal both result in an up-core "smearing" of a charcoal peak derived from a single fire (Gardner and Whitlock 2001, Whitlock and Millspaugh 1996).

Table 7.2. Fire Proxy Autocorrelations
Correlations of Charcoal Influx with Future Values of Itself

Significant Autocorrelations

# of Lags	Correlation (Pearson's r)	Significance (p)
1 lag (50 years)	+0.763	< 0.0001
2 lags (100 years)	+0.615	< 0.007

Insignificant Autocorrelations

# of Lags	Correlation (Pearson's r)	Significance (p)
3 lags (150 years)	+0.536	< 0.027
4 lags (200 years)	+0.435	< 0.092
5 lags (250 years)	+0.323	< 0.241
6 lags (300 years)	+0.072	< 0.806

Summary of Charcoal Results

The record of charcoal particles from local and extralocal fire sources stretches back to 1,525 B.P. at Five Lakes. This proxy record shows that fire events occurred every 148 years on average, with no clear trends or periods of increased or decreased fire frequency. Also, charcoal influx is greatest from 1,525 B.P to about 500 B.P, drops around 500 B.P. and remains low until the present. The cause of this is unclear, and could be due to a change in the type, size, or distance of the fire of origin. Charcoal values are also highly positively autocorrelated in the short term (50-100 years), but this seems to be a function of the sedimentary processes at work rather than a function of the source fires producing the charcoal.

Putting Five Lakes Charcoal Results in Context

How do the return intervals obtained at Five Lakes compare with other records of fire frequency? The higher elevations in the study area currently support pine forests, and pine has been a dominant taxon throughout the span of the record being examined. Historical records and fire scar studies in the Northwest provide information on local processes. At this scale, pine forests tend to have return intervals of 6 to 47 years between fires (Hall 1976; Langston 1999; Maruoka 1994; Soeriaatmadja 1966; Weaver 1959, 1961, 1967); and return intervals of 2 to 25 years are reported specifically for ponderosa pine (Fulé et al. 1997, 2003; Baker and Ehle 2001). In contrast, palynological studies from forested areas of western North America such as Washington, Oregon, Idaho, and Wyoming indicate that longer return intervals of 40 to 283 years are recognizable in charcoal records (Chatters 1998, Chatters and Leavell 1995, Millspaugh and Whitlock 1995). This indicates that particle proxy records can be used to sense past fire events, but that the events uncovered from studies of lacustrine sediments record only those occurring on a larger scale than events revealed through fire scar studies.

The data from Five Lakes, indicating return intervals averaging 148 years, and ranging from 94 to 232 years is perfectly in line with expectations from palynological data. This interval is longer than one would expect based on historical and fire scar records, which is understandable given the nature and spatial and temporal scale sampled by each record. It is probable that charcoal proxies are merely less sensitive than historical and fire scar records—charcoal seems to only record larger, more infrequent events. In comparison to other charcoal records, Five Lakes seems to record extra-local fire frequencies that are normal for prehistoric proxy records throughout the western United States. In other analyses of charcoal records from the interior Northwest, authors have tended to interpret shorter return intervals (40-90 years) as evidence of anthropogenic fire regimes (Chatters 1998a). If this is the case, then the longer return rates (averaging 148 years) seen at Five Lakes might be attributable to natural and not cultural fires.

Does fire, as recorded by charcoal, show a relationship with climate? Return intervals between fires do not change throughout the record, but climate does (Chapter 5). This is surprising, but other researchers have noted anomalous disjunctures between the expected relationship between climate and fire frequency. For example, Bergeron and Archambault (1993) expected fire frequency to increase during the warming that took place after the end of the LIA, but fire frequency decreased instead.

Although fire frequency seems to be unrelated to climate, there is an interesting potential relationship between fire amplitude (magnitude of charcoal the influx) and local aridity. Oxygen isotope records from Hidden and Rinker's Lakes record a period of relatively warm, dry conditions starting 600 B.P. and reaching maximum aridity around 400 B.P. (Figure 5.2). Sometime between 600 and 500 B.P. when values of *Pinus* pollen dropped, disturbance taxa increased (Figure 6.1), and the influx of large charcoal particles plummeted (Figure 7.3). This coincident change in several records gives tantalizing clues pointing to a potential relationship between fire frequency, climate, and vegetation. This is curious since palynological studies from the Plateau and the research area indicate a general trend over the Holocene in which xeric conditions promoted the growth of pines and grasses, two taxa that thrive under conditions of frequent low-intensity fires. It could be that local decreases in available moisture after 600 B.P. caused a decrease in pines, and that this resulted in less fuel available for fires. However, as plots of human population (Figures 3.1 through 3.3) show, the number of people in the study area declined after 550 years ago as well. This similar timing to the suite of other events provides a complicating factor that can only be evaluated statistically, through correlations, multivariate partitioning, and an analysis of repeated lead-lag relationships over time.

Chapter 8. Results and Discussion

A major product of this study is a database, consisting of comparable proxies for fire, human population, vegetation, and climate, sampled at 50-year intervals from A.D. 950 to 1900. These data are listed in Appendix H and provide an invaluable resource for investigating questions about the long-term dynamics of human ecology. Elements of this database were either generated by this study (e.g., the charcoal, pollen, and isotope data) or drawn from earlier work in the area (the population proxy). To make all variables comparable, values are given for each variable at the same 50-year intervals over the same time span as the pre-existing population proxy (Campbell 1989, 1990). This ensures that values from the same interval are coeval, and ensures that sampling is done for each variable at the same resolution. To accomplish this, mathematical interpolation between existing data points was required for some proxies to match temporal scales between variables and provide sampling at equal intervals over time (allowing for such statistical analyses as time series). The resulting database provides the basis for the statistical analyses provided in this chapter, and allows for the investigation of a wide range of questions about the interaction and interrelationships in the landscape over time at different time scales.

Population was entered into the analysis using a population index that is a composite of all of Campbell's (1989, 1990) proxy population variables (as presented in Chapter 3). The climate proxy used here is also a combination of two records—oxygen isotope records from Hidden and Rinker's Lakes. The population and isotope indexes were constructed in analogous manners, by standardizing and compiling separate records into a single synthetic product. For each, the individual sets of values were re-expressed as standardized deviations from a sample mean, and these Z-scores were added together for each successive time period. This produced an average score for each set of observations for each time period, as described and presented in Chapters 3 and 5.

The vegetation proxy used in this chapter was the percent composition of different taxa from pollen drawn from Five Lakes sediments. The proxy used for fire was the absolute influx ($\#/cm^2/yr$) of large (125-500 μm diameter) charcoal particles into Five Lakes. All these proxies are listed in Appendix H, and used for running correlations and time series tests to better understand the interactions between these variables.

Autocorrelations were run on each variable, comparing it to future values of itself. Successive correlations were run with SPSS v.10, introducing an additional 50-year (one interval) lag with each iteration. This was a simple way to determine if a variable was autocorrelated, or dependent upon past values of itself. Many autocorrelations were expected to be found since the factors being studied—human population, fire frequency, climate, and vegetation—are logically conditioned by

their own history. Although autocorrelations were already presented earlier, they will be reviewed here as well.

In addition to autocorrelations, bivariate correlations were run between different variables for coeval sets of values (again using SPSS V.10). This helped to determine, for example, if contemporaneous values of human population and fire were linked with one another. In order to better assess the potential cause-effect relationships between variables, bivariate correlations were run between variables again once a time lag was introduced in the model (using SPSS v.10). High correlations between a leading proxy and lagging proxy provide suggestive evidence about the response of the system and possible causal relationships. For example, a leading population proxy positively correlated with a lagging fire proxy could indicate that the increased presence of humans somehow causes an increase in the amount of fuel being consumed by fire on the landscape.

The logical last step in the statistical analysis was to evaluate if the correlations seen in the lagged univariate and bivariate analyses had explanatory value in a multivariate system. To understand the interrelationships of all factors, simultaneously, time series analysis (in the time domain) was run on the database given in Appendix H (using the eViews V.4—a statistical software package used primarily by economists). Time series analysis allows multiple correlations, autocorrelations, and linear independence of many variables to be considered simultaneously, while successively introducing lags of varying lengths into the model. Time series in the time domain allows for the effects of one variable, such as human population size, to be isolated and examined on various time scales. As such, it can help determine the relative contribution of anthropogenic inputs into the system. For example, one can ask if human population, and human population alone, adequately predicts future fire. Determining if human action is both significantly and linearly independent from other potential explanatory factors requires a more sophisticated procedure than simple bivariate correlations.

Granger causality, originally developed for two-variable systems (Gottman 1981, Granger 1969), but now able to be used in multivariate settings, was used for this analysis. A leading variable is said to "Granger cause" a lagging variable if the leading variable is: (1) significantly correlated with the lagging variable, (2) linearly independent of other competing independent variables, and (3) not autocorrelated with itself. That is, Granger causality helps to identify those variables that are acting as independent forcing agents in a system. Using such statistical tools as time series and bivariate correlation, especially in combination, the long-term interactions of ecological systems can be examined in depth (Delcourt and Delcourt 1991, Green 1981). The discussion, presented below, looks at the potential factors influencing each variable, in turn, starting with fire. Results for each factor begin with simple models of bivariate correlation, and then proceed to more complex analyses involving lead-lag relationships and assessments of the differential contribution of factors within a multivariate analysis. For all the statistical procedures used, a *p*-value of 0.01 or less was considered to be significant.

What Factors Influence the Fire Proxy?

Is Fire Statistically Influenced by Humans?

It is assumed by many researchers that humans increase fire frequency on a landscape primarily by increasing the number of ignition events that could start vegetation fires (Denevan 1992; Kay 2000, 2007; Parker 2002b; Pyne 1982; Vale 2002a), especially at low population sizes (Guyette and Dey 2000; Guyette and Spetich 2003; Guyette et al. 2002, 2006). If human presence and fire in eastern Washington are linked, then increases in human population should have a significant positive correlation with the deposition of large charcoal particles into area lakes. Correlations between contemporaneous values of charcoal influx and human population (see Table 8.1) support this idea, as they are significantly correlated at the 0.01 level ($r = +0.771$ $p \leq 0.0001$).

Although coeval proxies of fire and human population are significantly correlated, it is even more informative to introduce leads and lags between the two variables and rerun bivariate correlations. If the two are statistically significantly correlated once lags are introduced, the leading (and likely causal) factor can be identified. This can reveal if changes in population lead to changes in fire. In order to investigate this, a series of bivariate correlations were run with either fire or human proxies in the lead. The first pair was run with a lag of 50 years, the next with a lag of 100 years, then 150 years, etc.

The sets of increasingly lagged bivariate correlations run on the fire proxy (large particle charcoal influx) and the human population index indicate that the two are interrelated, and in the manner assumed by paleoecologists. Results from this lead-lag type of analysis (given in Table 8.2) show that when population is allowed to assume the lead role, significant correlations between population and later values of charcoal (fire) are seen. Significant positive correlations are seen when lags of one interval (50 years), two intervals (100 years), or three intervals (150 years) are introduced into the model. Whether examined synchronically or diachronically, population and fire show a strong interrelationship in bivariate models, with fire correlated with population size up to 150 years in the past. This is in keeping with other quantitative statistical paleoecological investigations, which have demonstrated a significant positive association between Native American population sizes and increased outbreak of fire (Black et al. 2006; Guyette and Dey 2000; Guyette and Spetich 2003; Guyette et al. 2002, 2006).

Table 8.1. Bivariate Correlations between Contemporary Values of Population, Oxygen Isotopes, and Charcoal

Variable Pairs	Correlation (Pearson's *r*)	Significance (*p*)
Charcoal-Oxygen Isotope	+0.110	< 0.645
Population-Oxygen Isotope	+0.308	< 0.187
Charcoal-Population	+0.772	< 0.0001

Table 8.2. Bivariate Correlations between Population (Leading) and Lagged Charcoal

Significant Correlations		
# of Lags	Correlation (Pearson's *r*)	Significance (*p*)
1 lag (50 years)	+0.840	< 0.0001
2 lags (100 years)	+0.713	< 0.001
3 lags (150 years)	+0.614	< 0.009

Insignificant Correlations		
# of Lags	Correlation (Pearson's *r*)	Significance (*p*)
4 lags (200 years)	+0.533	< 0.033
5 lags (250 years)	+0.394	< 0.147
6 lags (300 years)	+0.312	< 0.277

Does Past Fire History Statistically Influence Future Fire Regimes?

Researchers believe that fire frequency in a given area is partially dependent upon the extent and frequency of past fires in an area. Vegetation fires should deplete fuels available on the landscape, and lead to a decreased chance of fires in the near future (Martin et al. 1977). A lack of fire should cause a buildup of fuel and increase the likelihood or "danger" of fire (Rauw 1980). The use of prescribed burns to prevent intense and damaging forest fires, and the contribution of historic fire suppression activities to the devastating Yellowstone fires of 1988 are examples of this (Romme and Despain 1989). Based on this scenario, the charcoal proxy for fire should be significantly negatively correlated with future fire.

This question relies, again, on an approach that introduces lags into a model that attempts to correlate values of charcoal influx from different times in the study area. The results of an autocorrelation analysis, run on the charcoal fire proxy, were presented earlier, and are listed in Table 7.2. As expected, there are significant correlations at the 0.01 level between charcoal and future charcoal values when one or two lags (50-100 year lags) are entered into the model. As the lags get longer and time more distant, the magnitude and significance of the relationship decrease.

Contrary to expectations, the significant correlations between charcoal and future charcoal values are positive correlations, not negative ones. Charcoal influx during one time period is very similar to influx during the next period. This could be because the incidence of fire increases the likelihood of fire in the near future on the landscape. It could also be a purely mechanical or stratigraphic problem inherent in the proxy itself. Charcoal in one sample could be diffused or "smeared" a short distance up the stratigraphic column by limited sediment mixing. Small scale bioturbation within the lake sediment sequence could bring some charcoal particles upwards in the stratigraphy. Also, years of reworking and redeposition of charcoal originally deposited the uplands could explain the observed positive autocorrelation in this variable (Clarke 1987).

Does Local Climate Influence Fire Frequency?

Climate is yet another factor assumed to influence the fire history of an area. In many areas, warm dry years promote the buildup of fuel, the outbreak of fire, and the spread of established fires (Fulé et al. 2003, Hall 1989). Is climate seen as having any effect on fire in the proxy records from the past 1,000 years in the study area? Comparing the contemporaneous values of the charcoal fire proxy and oxygen isotope climate proxy (Table 8.1), the two are positively but not significantly correlated with one another ($r = +0.110$, $p < 0.645$). Climate, as measured by oxygen isotope values (which, in turn, reflect relative aridity) is not related to the coeval incidence of fire at a 0.01 significance level.

Perhaps climate influences fire, but that influence takes time to manifest itself in the record. This could easily happen if climatic conditions took several years to influence the vegetative growth of plants already growing on the landscape, or if it took many years to influence the abundance of different plant taxa on the landscape. Vegetation change would then affect the type and amount of fuel available on the landscape, and that, in turn, could pattern fire regimes. With such mechanisms at work, it is perfectly reasonable to assume that fire response to climatic conditions could take years, decades, or centuries to manifest itself. Dendrochronological studies that include fire scars show that fires are more likely to occur in dry years that follow close after a series of wetter-than-average years, since wet years promote the growth of vegetation and arid years dry out this new fuel, making the landscape more flammable (Fulé et al. 2003, Guyette et al. 2006).

Does fire respond, but with a lag, to climate? A series of lagged bivariate correlation analyses relating climate and fire proxies is presented in Table 8.3. According to the statistical results, charcoal values are significantly related to prior oxygen isotope values when 5 or 6 lags (250 and 300 years) are introduced into the model. This relationship is statistically significant and positive, indicating that times of more positive oxygen isotope values, indicative of arid conditions, lead to times of increased deposition of large charcoal particles as evidence of resulting local fires some 250 to 300 years in the future.

Do Abundances of Different Plant Taxa Influence the Fire Proxy?

The outbreak and spread of fire should be related to the available fuels on the landscape (related to the amount of fuel, its moisture content, texture, and the type or taxonomic origin of the dead material), as well as to the presence of an ignition source (humans, lightning, etc). Modern ecological studies have shown that the type of vegetation in an area influences the outbreak and spread of fire (Marouka 1994). Grasses, sagebrush, and ponderosa pine, for instance, all produce different kinds of dead tissues, with different characteristics as a fuel for combustion. The structure, particle size, chemical content, and density of materials (Ellis 1980) are all dependent on the taxa providing the fuel and the packing of individual plants together. It stands to reason that the vegetation variables provided by the pollen analysis would have some relation to charcoal influx as a measure of local fires.

The role of vegetation can be evaluated by first comparing contemporaneous values of different plant taxa with the influx of large charcoal particles. Table 8.4 lists the significant bivariate correlations between plant taxa from Five Lakes and contemporaneous charcoal values (with insignificant correlations given in Table 8.5). Again, rare types are not included, as the abundances of these categories can not be reliably estimated from the 500-grain pollen counts. The

Table 8.3. Bivariate Correlations between Oxygen Isotope (Leading) and Lagged Charcoal Influx

	Significant Correlations	
# of Lags	Correlation (Pearson's *r*)	Significance (*p*)
5 lags (250 years)	+0.704	< 0.003
6 lags (300 years)	+0.791	< 0.001

	Insignificant Correlations	
# of Lags	Correlation (Pearson's *r*)	Significance (*p*)
1 lag (50 years)	+0.218	< 0.371
2 lags (100 years)	+0.240	< 0.337
3 lags (150 years)	+0.376	< 0.137
4 lags (200 years)	+0.586	< 0.017

remaining major and minor taxa appear in normal typeface in the tables in this chapter, with aggregate categories like NAP and AP in bold typeface.

Of the taxa listed as statistically significant, *Pinus* is the only one that has a statistically significant positive correlation with contemporaneous values of the fire proxy at the 0.01 level. The percent abundance of pine is tightly correlated with the amount of large charcoal particles in the same samples, with an associated Pearson's $r = +0.783$ at a significance of $p \leq 0.0001$. Of the remaining reliable minor taxa, several are significantly negatively correlated with the fire proxy; *Betula*, Cheopodiaceae, *Artemesia*, *Ambrosia*, and *Aster* all show a negative relationship (all with *r* ranging from -0.723 to -0.770 and all with $p \leq 0.0001$). All non-arboreal taxa (NAP), taken together, also show a statistically significant inverse relationship with contemporaneous values of charcoal influx (this relationship has $r = -0.764$ and $p \leq 0.0001$).

Times of high pine domination, then, correlate with outbreaks of fire in the area. Either ponderosa pine-dominated vegetation encourages the outbreak and spread of fire, or the same conditions that promote the growth and dominance of pine trees also promote the outbreak of fires. It is highly unlikely that the statistical correlation indicates that fire promotes pines since each sample represents approximately a ten-year period of deposition and it takes longer than 10 years for pines to establish new seedlings and for these new trees to mature. Whatever the mechanism of causation behind this statistical relationship, fire has been more common in pine-dominated vegetation by far than in any other type of vegetation over the past thousand years in this area. Likewise, fire has been rare in times when non-arboreal taxa have been relatively common; fire has been unlikely on the landscape whenever taxa like Chenopodiaceae,

Betula, and Asteraceae (*Artemesia*, *Ambrosia*, *Aster* and Ligulaflorae) are enjoying times of relative abundance.

How do these relationships change once lags are introduced into the model? Do certain taxa or vegetation types lead to future fires? Lagged bivariate correlations between the charcoal variable and the pollen from different taxa reveal significant positive and negative bivariate relationships as summarized in Table 8.6 (insignificant correlations are listed in Table 8.7). As Table 8.6 indicates, episodes of fire can be statistically predicted in bivariate models from the values of various taxa during previous time periods. Statistically significant positive relationships exist between the percentage of *Pinus* pollen in samples and the influx of charcoal at 50, 100, 150, or 200 years later. Unlike *Pinus*, *Picea* is not significantly correlated with contemporaneous charcoal values, yet *Picea* shows a positive relationship with charcoal influx when greater time depth is added. As Table 8.6 indicates, *Picea* is significantly correlated with charcoal at 4, 5, or 6 lags (200, 250, or 300 years).

Significant negative relationships exist for many of the minor taxa, primarily for *Betula*, Chenopodiaceae, Poaceae, and Asteraceae (*Aster*, *Ambrosia*, *Artemesia*, and Ligulaflorae). The significant relationships in Table 8.6 hold for several successive lags; significant negative correlations occur for *Betula* at 1, 2, and 3 lags, for Poaceae at 2, 3 and 4 lags, for Chenopodiaceae at 1 and 2 lags, and for members of the Asteraceae family at 1, 2, 3, 4, 5, and 6 lags. Most of these significant leading taxa (all but Poaceae) were also significantly negatively correlated with contemporaneous charcoal values (Table 8.4). This indicates that the relationships seen between contemporary variables also occur between past vegetation and fire as well. In a bivariate model, fire is associated with high percentages of past and present pine trees, and fire is inversely related with the abundance of typical "disturbance" species such as Chenopodiaceae, *Betula*, and Asteraceae.

Table 8.4. Significant Bivariate Correlations between the Percentage of Pollen Types and Contemporaneous Charcoal Influx

Positive Correlations

Pollen Type	Correlation (Pearson's *r*)	Significance (*p*)
Pinus	+0.783	< 0.0001
AP	**+0.764**	**< 0.0001**

Negative Correlations

Pollen Type	Correlation (Pearson's *r*)	Significance (*p*)
Betula	-0.770	< 0.0001
Chenopodiaceae	-0.748	< 0.0001
Artemesia	-0.723	< 0.0001
Ambrosia	-0.740	< 0.0001
Aster	-0.768	< 0.0001
NAP	**-0.743**	**< 0.0001**

Results are for major and minor taxa, only. Items in bold are synthetic categories (NAP and AP).

Table 8.5. Insignificant Bivariate Correlations between the Percentage of Pollen Types and Contemporaneous Charcoal Influx

Positive Correlations

Pollen Type	Correlation (Pearson's *r*)	Significance (*p*)
Picea	+0.176	< 0.458

Negative Correlations

Pollen Type	Correlation (Pearson's *r*)	Significance (*p*)
Alnus	-0.437	< 0.054
Poaceae	-0.419	< 0.066
Ligulaflorae	-0.557	< 0.011

Which Variables Influence Fire Most, and How do They Interact?

Contemporaneous and lagged bivariate analyses have helped to identify the variables that relate to the influx of charcoal as a fire proxy. Identified as significant influences in bivariate tests are: prior human population levels (with a 0, 50, 100, and 150 year lead), prior oxygen isotope values (with a 250 or 300 year lead), and the present and past abundances of taxa like Chenopodiaceae, *Pinus*, *Betula*, and Asteraceae.

Which of these variables are most important in determining fire frequency? Answering this question requires a simultaneous comparison of the various inputs, best addressed with multivariate statistics. Granger causality is a multivariate technique that can be used to assess the unique contribution each conditioning variable makes to the fire proxy. Granger causality identifies those leading factors that add to a predictive model, taking other independent variables as well as the dependent variable (and its own feedback potential) into account. Like multiple regression, Granger causality requires that a given variable, X, be correlated with the predicted variable Y, and that X not be highly correlated with other potential explanatory variables. Unlike multiple regression, however, Granger causality also requires that past values of both Y and X not be correlated with X. This procedures eliminates problems of autocorrelation and correlations between the independent "explanatory" variables (Gottman 1981). When tests of Granger causality are run on the potential predictor variables identified as significant in the bivariate tests (in Tables 8.1, 8.2, 8.3, 8.4, and 8.6), several variables are identified as having statistically significant predictive "causal" value.

Table 8.6. Significant Bivariate Correlations between the Percentage of Pollen Types (Leading) and Lagged Charcoal Influx

Positive Correlations

Pollen Type	# Lags	Years	Correlation (Pearson's *r*)	Significance (2-tailed)
Pinus	1	50	+0.841	< 0.0001
AP	**1**	**50**	**+0.833**	**< 0.0001**
Pinus	2	100	+0.858	< 0.0001
AP	**2**	**100**	**+0.875**	**< 0.0001**
Pinus	3	150	+0.829	< 0.0001
Picea	3	150	+0.638	< 0.006
AP	**3**	**150**	**+0.866**	**< 0.0001**
Pinus	4	200	+0.758	< 0.001
Picea	4	200	+0.751	< 0.001
AP	**4**	**200**	**+0.810**	**< 0.0001**
Picea	5	250	+0.832	< 0.0001
Picea	6	300	+0.817	< 0.0001

Negative Correlations

Pollen Type	# Lags	Years	Correlation (Pearson's *r*)	Significance (2-tailed)
Betula	1	50	-0.776	< 0.0001
Chenopodiaceae	1	50	-0.684	< 0.001
Artemesia	1	50	-0.817	< 0.0001
Ambrosia	1	50	-0.790	< 0.0001
Aster	1	50	-0.759	< 0.0001
NAP	**1**	**50**	**-0.833**	**< 0.0001**
Betula	2	100	-0.764	< 0.0001
Chenopodiaceae	2	100	-0.700	< 0.001
Poaceae	2	100	-0.757	< 0.0001
Artemesia	2	100	-0.832	< 0.0001
Ambrosia	2	100	-0.796	< 0.0001
Aster	2	100	-0.676	< 0.002
NAP	**2**	**100**	**-0.875**	**< 0.0001**
Betula	3	150	-0.682	< 0.003
Poaceae	3	150	-0.862	< 0.0001
Artemesia	3	150	-0.838	< 0.0001
Ambrosia	3	150	-0.764	< 0.0001
Aster	3	150	-0.673	< 0.003
NAP	**3**	**150**	**-0.866**	**< 0.0001**
Poaceae	4	200	-0.787	< 0.0001
Artemesia	4	200	-0.847	< 0.0001
Ambrosia	4	200	-0.723	< 0.002
Aster	4	200	-0.709	< 0.003
NAP	**4**	**200**	**-0.810**	**< 0.0001**
Artemesia	5	250	-0.752	< 0.001
Aster	5	250	-0.665	< 0.007
NAP	**5**	**250**	**-0.641**	**< 0.010**
Artemesia	6	300	-0.694	< 0.006

**Table 8.7. Insignificant Bivariate Correlations between the Percentage of
Pollen Types (Leading) and Lagged Charcoal Influx**

Positive Correlations

Pollen Type	# Lags	Years	Correlation (Pearson's *r*)	Significance (2-tailed)
Picea	1	50	+0.417	< 0.076
Picea	2	100	+0.513	< 0.029
Pinus	5	250	+0.612	< 0.015
Pinus	6	300	+0.500	< 0.069
Chenopodiaceae	6	300	+0.104	< 0.724
AP	**6**	**300**	**+0.542**	**< 0.045**

Negative Correlations

Pollen Type	# Lags	Years	Correlation (Pearson's *r*)	Significance (2-tailed)
Alnus	1	50	-0.478	< 0.038
Poaceae	1	50	-0.563	< 0.012
Ligulaflorae	1	50	-0.565	< 0.012
Alnus	2	100	-0.521	< 0.027
Polygonaceae	2	100	-0.408	< 0.093
Ligulaflorae	2	100	-0.475	< 0.047
Alnus	3	150	-0.573	< 0.016
Chenopodiaceae	3	150	-0.590	< 0.013
Ligulaflorae	3	150	-0.108	< 0.679
Chenopodiaceae	4	200	-0.414	< 0.111
Polygonaceae	4	200	-0.431	< 0.095
Betula	5	250	-0.583	< 0.023
Alnus	5	250	-0.347	< 0.205
Chenopodiaceae	5	250	-0.176	< 0.531
Poaceae	5	250	-0.551	< 0.033
Ambrosia	5	250	-0.629	< 0.012
Polygonaceae	5	250	-0.246	< 0.376
Ligulaflorae	5	250	-0.044	< 0.875
Betula	6	300	-0.398	< 0.159
Alnus	6	300	-0.271	< 0.350
Poaceae	6	300	-0.431	< 0.124
Ambrosia	6	300	-0.529	< 0.052
Aster	6	300	-0.591	< 0.026
Polygonaceae	6	300	-0.203	< 0.486
Ligulaflorae	6	300	-0.039	< 0.895
NAP	**6**	**300**	**-0.542**	**< 0.045**

Table 8.8 lists all leading variables that pass the test for Granger causality (based on raw output from Scharf 2002). To simplify the list somewhat, the list of interactions is narrowed to just those involving oxygen isotopes, charcoal, and/or population and this shorter list is presented in Table 8.9. Table 8.10 lists only the significant Granger causal relationships that involve fire. Leading fire by fifty years are values of *Pinus* and human population. Also identified as important via Granger causality (Table 8.10) is the oxygen isotope index 300 years prior to the outbreak of fire. Although the bivariate tests had indicated that *Pinus*, population, and oxygen isotopes were significant at other lag-intervals as well, once the influence and interaction of other variables such as taxonomic abundances were taken into account, only the fifty year lags remain as significant. Granger causality, then, removes those variables that suffer from multicollinearity and helps solve the problem of equifinality.

Again, this method identifies past climate, *Pinus* (the most abundant and ubiquitous taxon), and human population as the primary determinants of the incidence of fire. Based on this information, paired with the sign of the bivariate correlation coefficient (Table 8.10), fire is most likely to occur 300 years after an arid period. Fire also significantly increases 50 years after plant associations high in pine and low in herbs and shrubs, and 50 years after a rise in human population. When multiple variables are examined simultaneously, then, patterning of fire is not determined by solely by one factor (such as human population), but by a combination of factors interacting on several different time frames.

What Factors Influence Vegetation?

The factors influencing vegetation can be examined using the same logic as outlined above. Bivariate analyses followed by multivariate evaluations identify and tease apart interrelationships within the dataset. This, in turn, clarifies the interactions between variables in the system over time.

Is Vegetation Influenced by Humans?

If humans cause vegetation change, it is logical to assume that the greater the number of humans on the landscape, the greater their impact on the landscape and its vegetation. Also central to this argument is that human activity will favor seral or "disturbance" vegetation, and cause a reduction in species that are typical of later-successional vegetation stages. This can be tested by comparing the human population proxy with the changing taxonomic composition of the pollen record. If human presence leads to disturbance and clearance, then the population proxy should be significantly and positively correlated with the classic disturbance species—*Alnus*, *Betula*, and weedy herbaceous taxa like Chenopodiaceae, Poaceae, and Asteraceae.

Table 8.11 shows the significant bivariate correlations between the population proxy and coeval pollen types

(Table 8.12 lists the insignificant correlations). The proxy records do not show a significant positive relationship between human population levels and classic "disturbance" taxa such as Poaceae, Chenopodiaceae, Asteraceae, *Betula*, and *Alnus*. In fact, surprisingly, there is a significant and strong negative relationship between humans and "disturbance" taxa. In contrast, the dominant and climax taxon (*Pinus*) is significantly positively correlated with contemporaneous values of human population, which could be construed as evidence against humans acting as disturbance agents in this case. This is even more curious when the archaeobotanical data from the area are considered; in the study area conifer remains collected as firewood are common in archaeological samples, with significant collection and consumption of ponderosa pine, fir, larch, and Douglas fir by people in prehistoric times (Stenholm 1985).

What could account for this set of relationships? Perhaps human action does, as predicted, result in the increase of "weedy," early-seral vegetation, but it takes time for the effects to manifest themselves. To explore this possibility, lags can be introduced and correlations run for different time scales. Table 8.13 summarizes the significant results of such lagged bivariate models involving plant abundances and human population size (with insignificant results listed in Table 8.14).

When human population size is allowed to be in a lead position, many of the same relationships as seen between contemporaneous values are repeated. Again, human population is significantly positively related to values of climax taxa and inversely related to seral vegetation. Human population is significantly negatively correlated with future amounts of herbaceous taxa like Chenopodiaceae (at lags of 50-150 years) or Asteraceae (at lags of 50 to 150 years), and with future abundances of early successional tree types (like *Alnus* and *Betula*) at longer lags. Human population in a bivariate model, then, is correlated significantly and negatively with both contemporaneous and future values of disturbance taxa, in direct conflict with the expectation that human action would encourage the growth of seral vegetation.

Does Fire Influence the Taxa on the Landscape?

Given that human population does not have the expected effect on vegetation, what does pattern future vegetation? Does the dataset reveal any relationships between fire and resulting plant communities? Again, interrelationships can be explored by first examining the bivariate relationships between fire and different plant taxa. Correlations between contemporaneous values of charcoal and the abundances of different kinds of plants indicates that fire is significantly positively correlated with coeval values of *Pinus* and significantly negatively correlated with coeval values of seral taxa (such as *Betula*, Chenopodiaceae, and Asteraceae).

Next, the fire proxy was put into a leading role in a series of bivariate correlations over different time scales. The significant results are listed in Table 8.15 (and the

**Table 8.8. List of All Significant Multivariate Granger Causal Factors
(Including Relationships between Taxa)**

Leading "Causal" Variable		Lagging "Responding" Variable	# of Lags	Lag (years)
Artemesia	→	*Betula*	1	50
Artemesia	→	Chenopodiaceae	1	50
Artemesia	→	*Pinus*	1	50
Aster	→	Chenopodiaceae	1	50
Aster	→	Ligulaflorae	1	50
population	→	*Betula*	1	50
Pinus	→	*Betula*	1	50
charcoal	→	Chenopodiaceae	1	50
population	→	charcoal	1	50
Pinus	→	charcoal	1	50
oxygen isotope	→	*Picea*	1	50
Alnus	→	population	2	100
Chenopodiaceae	→	*Ambrosia*	2	100
Ambrosia	→	Chenopodiaceae	2	100
Artemesia	→	Ligulaflorae	2	100
Aster	→	Chenopodiaceae	2	100
oxygen isotope	→	*Aster*	2	100
Picea	→	*Aster*	2	100
Betula	→	Ligulaflorae	2	100
Pinus	→	*Betula*	2	100
population	→	Chenopodiaceae	2	100
Pinus	→	Ligulaflorae	2	100
Alnus	→	population	3	150
Artemesia	→	*Pinus*	3	150
Aster	→	Chenopodiaceae	3	150
Alnus	→	population	4	200
Artemesia	→	*Ambrosia*	4	200
Ambrosia	→	Ligulaflorae	4	200
Aster	→	Chenopodiaceae	4	200
charcoal	→	population	4	200
oxygen isotope	→	Chenopodiaceae	4	200
Chenopodiaceae	→	Ligulaflorae	5	250
oxygen isotope	→	Chenopodiaceae	5	250
Chenopodiaceae	→	*Pinus*	5	250
Pinus	→	*Picea*	5	250
oxygen isotope	→	charcoal	6	300

Table 8.9. Granger Causality—All Significant Results Involving Population, Charcoal, and/or Oxygen Isotope Values

Leading "Causal" Variable		Lagging "Responding" Variable	# of Lags	Lag (years)
population	→	*Betula*	1	50
charcoal	→	Chenopodiaceae	1	50
population	→	charcoal	1	50
Pinus	→	charcoal	1	50
oxygen isotope	→	*Picea*	1	50
	→			
	→			
Alnus	→	population	2	100
oxygen isotope	→	*Aster*	2	100
population	→	Chenopodiaceae	2	100
	→			
Alnus	→	population	3	150
	→			
Alnus	→	population	4	200
charcoal	→	population	4	200
oxygen isotope	→	Chenopodiaceae	4	200
	→			
oxygen isotope	→	Chenopodiaceae	5	250
	→			
oxygen isotope	→	charcoal	6	300

Table 8.10. List of Significant Multivariate Granger Causal Factors Involving Charcoal, with Corresponding Lagged Bivariate Correlations

Charcoal in Leading Position

Leading "Causal" Variable	Lagging "Responding" Variable	# of Lags	Lag (years)	Bivariate Correlation Pearson's *r*	Significance of Correlation
charcoal → Chenopodiaceae		1	50	-0.823	< 0.0001
charcoal → population		4	200	+0.791	< 0.001

Charcoal in Lagging Position

Leading "Causal" Variable	Lagging "Responding" Variable	# of Lags	Lag (years)	Bivariate Correlation Pearson's *r*	Significance of Correlation
population → charcoal		1	50	+0.840	< 0.0001
Pinus → charcoal		1	50	+0.841	< 0.0001
oxygen isotope → charcoal		6	300	+0.791	< 0.001

Table 8.11. Significant Bivariate Correlations between Pollen Percentages and Human Population

Positive Correlations

Pollen Type	Correlation (Pearson's *r*)	Significance (*p*)
Pinus	+0.868	< 0.0001
AP	**+0.881**	**< 0.0001**

Negative Correlations

Pollen Type	Correlation (Pearson's *r*)	Significance (*p*)
Betula	-0.780	< 0.0001
Alnus	-0.581	< 0.007
Chenopodiaceae	-0.694	< 0.001
Poaceae	-0.691	< 0.001
Artemesia	-0.887	< 0.0001
Ambrosia	-0.681	< 0.001
Aster	-0.788	< 0.0001
Ligulaflorae	-0.598	< 0.005
NAP	**-0.881**	**< 0.0001**

Table 8.12. Insignificant Bivariate Correlations between Pollen Percentages and Human Population

Positive Correlations

Pollen Type	Correlation (Pearson's *r*)	Significance (*p*)
Picea	**+0.257**	**< 0.274**

*There are no insignificant negative correlations

insignificant ones in Table 8.16). Significant relationships seen over shorter time intervals are similar to those seen for contemporaneous values. As with the coeval correlations, *Pinus* is significant and positively correlated with fire, in a lead position at intervals of 50, 100, and 150 years. Asteraceae, *Betula*, and Chenopodiaceae, again, are significantly negatively correlated with charcoal, but the relationships disappear after 50 to 150 years (Table 8.15). The only surprise is *Alnus*, which is significantly negatively correlated with charcoal, but the relationship only shows up when longer lags (of 100 to 300 years) are introduced. In sum, increases in charcoal values correlate with coeval and future values of *Pinus*. This is in keeping with modern observations, which indicate that many species of pine are fire-dependent, not just fire-tolerant (Rauw 1980). The fire proxy is inversely related to most of the other taxa, as Chenopodiaceae, *Betula*, and Asteraceae are negatively correlated with fire on short time scales and with *Alnus* at longer time scales, indicating that burns discourage the future growth of these taxa.

Does Climate Influence Vegetation?

Given that there is a complex feedback between vegetation and fire in the study area over the past millennium, what is the role of climate in this system? Climate might be expected to respond to global processes and therefore to not be dependent on other factors. Compared to fire and vegetation, climate should be acting statistically independently when used as a lagged variable in bivariate statistical procedures. Climate should also be identified as a significant factor when multivariate time series are run and Granger causalities are evaluated.

It is logical to begin by examining the bivariate relationship in this system that has not yet been examined, the one between contemporaneous values of oxygen isotope values and taxonomic abundances. Tables 8.17 and 8.18 list the significant and insignificant correlations between these variables, respectively. Very few taxa are significantly correlated with contemporaneous oxygen isotope values. Three taxa—*Alnus*, Poaceae, and *Artemesia*—are negatively correlated with isotope values. Only one taxon, *Picea*, shows a significant and positive relationship. Based on these statistics, one would expect to find proportionately

Table 8.13. Significant Bivariate Correlations between Population (Leading) and Lagged Values of Taxonomic Abundance

Positive Correlations

Pollen Type	# Lags	Years	Correlation (Pearson's *r*)	Significance (2-tailed)
Pinus	1	50	+0.837	< 0.0001
AP	**1**	**50**	**+0.816**	**< 0.0001**
Pinus	2	100	+0.743	< 0.0001
AP	**2**	**100**	**+0.689**	**< 0.002**
Pinus	3	150	+0.735	< 0.001
AP	**3**	**150**	**+0.665**	**< 0.004**
Pinus	4	200	+0.674	< 0.004

Negative Correlations

Pollen Type	# Lags	Years	Correlation (Pearson's *r*)	Significance (2-tailed)
Betula	1	50	-0.828	< 0.0001
Chenopodiaceae	1	50	-0.795	< 0.0001
Artemesia	1	50	-0.757	< 0.0001
Ambrosia	1	50	-0.760	< 0.0001
Aster	1	50	-0.742	< 0.0001
NAP	**1**	**50**	**-0.816**	**< 0.0001**
Betula	2	100	-0.739	< 0.0001
Chenopodiaceae	2	100	-0.841	< 0.0001
Ambrosia	2	100	-0.817	< 0.0001
Aster	2	100	-0.616	< 0.006
NAP	**2**	**100**	**-0.689**	**< 0.002**
Betula	3	150	-0.746	< 0.001
Alnus	3	150	-0.603	< 0.010
Chenopodiaceae	3	150	-0.721	< 0.001
Ambrosia	3	150	-0.680	< 0.003
Liguiaflorae	3	150	-0.627	< 0.007
NAP	**3**	**150**	**-0.665**	**< 0.004**
Betula	4	200	-0.739	< 0.001
Alnus	4	200	-0.807	< 0.0001
Liguiaflorae	4	200	-0.620	< 0.010
Alnus	5	250	-0.889	< 0.0001
Abies	5	250	-0.676	< 0.006

Table 8.14. Insignificant Bivariate Correlations between Population (Leading) and Lagged Values of Taxonomic Abundance

Positive Correlations

Pollen Type	# Lags	Years	Correlation (Pearson's *r*)	Significance (2-tailed)
Picea	1	50	+0.120	< 0.626
Picea	2	100	+0.041	< 0.871
Picea	3	150	+0.150	< 0.566
Picea	4	200	+0.262	< 0.327
AP	**4**	**200**	**+0.584**	**< 0.017**
Pinus	5	250	+0.585	< 0.022
Picea	5	250	+0.335	< 0.222
Aster	5	250	+0.024	< 0.932
AP	**5**	**250**	**+0.512**	**< 0.051**
Pinus	6	300	+0.469	< 0.091
Picea	6	300	+0.409	< 0.147
Aster	6	300	+0.051	< 0.863
AP	**6**	**300**	**+0.441**	**< 0.114**

Negative Correlations

Pollen Type	# Lags	Years	Correlation (Pearson's *r*)	Significance (2-tailed)
Alnus	1	50	-0.478	< 0.039
Poaceae	1	50	-0.529	< 0.020
Ligulaflorae	1	50	-0.383	< 0.106
Alnus	2	100	-0.445	< 0.064
Poaceae	2	100	-0.407	< 0.094
Artemesia	2	100	-0.525	< 0.025
Ligulaflorae	2	100	-0.548	< 0.019
Poaceae	3	150	-0.504	< 0.039
Artemesia	3	150	-0.568	< 0.017
Aster	3	150	-0.432	< 0.083
Chenopodiaceae	4	200	-0.569	< 0.022
Poaceae	4	200	-0.561	< 0.024
Artemesia	4	200	-0.568	< 0.022
Ambrosia	4	200	-0.468	< 0.067
Aster	4	200	-0.164	< 0.544
NAP	**4**	**200**	**-0.584**	**< 0.017**
Picea	5	250	+0.335	< 0.222
Chenopodiaceae	5	250	-0.388	< 0.153
Artemesia	5	250	-0.531	< 0.041
Ambrosia	5	250	-0.311	< 0.259
Ligulaflorae	5	250	-0.606	< 0.017
NAP	**5**	**250**	**-0.512**	**< 0.051**
Betula	6	300	-0.455	< 0.102
Chenopodiaceae	6	300	-0.103	< 0.725
Poaceae	6	300	-0.658	< 0.010
Artemesia	6	300	-0.637	< 0.014
Ambrosia	6	300	-0.051	< 0.862
Ligulaflorae	6	300	-0.633	< 0.015
NAP	**6**	**300**	**-0.441**	**< 0.114**

**Table 8.15. Significant Bivariate Correlations between Charcoal (Leading)
and Lagged Values of Taxonomic Abundance**

Positive Correlations

Pollen Type	# Lags	Years	Correlation (Pearson's *r*)	Significance (2-tailed)
Pinus	1	50	+0.816	< 0.0001
AP	**1**	**50**	**+0.799**	**< 0.0001**
Pinus	2	100	+0.769	< 0.0001
AP	**2**	**100**	**+0.714**	**< 0.001**
Pinus	3	150	+0.660	< 0.004

Negative Correlations

Pollen Type	# Lags	Years	Correlation (Pearson's *r*)	Significance (2-tailed)
Betula	1	50	-0.736	< 0.0001
Chenopodiaceae	1	50	-0.823	< 0.0001
Artemesia	1	50	-0.698	< 0.001
Ambrosia	1	50	-0.753	< 0.0001
Aster	1	50	-0.696	< 0.001
NAP	**1**	**50**	**-0.799**	**< 0.0001**
Betula	2	100	-0.756	< 0.0001
Alnus	2	100	-0.626	< 0.005
Chenopodiaceae	2	100	-0.785	< 0.0001
Artemesia	2	100	-0.595	< 0.009
Ambrosia	2	100	-0.675	< 0.002
Ligulaflorae	2	100	-0.590	< 0.010
NAP	**2**	**100**	**-0.714**	**< 0.001**
Alnus	3	150	-0.633	< 0.006
Chenopodiaceae	3	150	-0.661	< 0.004
Ligulaflorae	3	150	-0.694	< 0.002
Alnus	4	200	-0.684	< 0.003
Poaceae	4	200	-0.441	< 0.087
Alnus	5	250	-0.653	< 0.008

Table 8.16. Insignificant Bivariate Correlations between Charcoal (Leading) and Lagged Values of Taxonomic Abundance

Positive Correlations

Pollen Type	# Lags	Years	Correlation (Pearson's *r*)	Significance (2-tailed)
Picea	1	50	+0.011	< 0.964
Picea	2	100	+0.020	< 0.939
Picea	3	150	+0.093	< 0.722
AP	**3**	**150**	**+0.577**	**< 0.015**
Pinus	4	200	+0.515	< 0.041
Picea	4	200	+0.173	< 0.521
AP	**4**	**200**	**+0.421**	**< 0.104**
Pinus	5	250	+0.363	< 0.184
Picea	5	250	+0.160	< 0.570
Aster	5	250	+0.125	< 0.658
AP	**5**	**250**	**+0.300**	**< 0.277**
Pinus	6	300	+0.338	< 0.237
Picea	6	300	+0.253	< 0.382
Ambrosia	6	300	+0.170	< 0.562
Aster	6	300	+0.148	< 0.613
AP	**6**	**300**	**+0.305**	**< 0.288**

Negative Correlations

Pollen Type	# Lags	Years	Correlation (Pearson's *r*)	Significance (2-tailed)
Alnus	1	50	-0.537	< 0.018
Poaceae	1	50	-0.562	< 0.012
Ligulaflorae	1	50	-0.519	< 0.023
Poaceae	2	100	-0.553	< 0.017
Artemesia	2	100	-0.595	< 0.009
Ambrosia	2	100	-0.675	< 0.002
Aster	2	100	-0.526	< 0.025
Alnus	3	150	-0.633	< 0.006
Poaceae	3	150	-0.465	< 0.060
Artemesia	3	150	-0.511	< 0.036
Ambrosia	3	150	-0.505	< 0.039
Aster	3	150	-0.245	< 0.343
NAP	**3**	**150**	**-0.577**	**< 0.015**
Betula	4	200	-0.586	< 0.017
Chenopodiaceae	4	200	-0.415	< 0.110
Poaceae	4	200	-0.441	< 0.087
Artemesia	4	200	-0.411	< 0.114
Ambrosia	4	200	-0.350	< 0.184
Aster	4	200	-0.010	< 0.970
Ligulaflorae	4	200	-0.435	< 0.092
NAP	**4**	**200**	**-0.421**	**< 0.104**
Betula	5	250	-0.379	< 0.164
Chenopodiaceae	5	250	-0.145	< 0.606
Poaceae	5	250	-0.432	< 0.108
Artemesia	5	250	-0.413	< 0.126
Ambrosia	5	250	-0.099	< 0.725
Ligulaflorae	5	250	-0.502	< 0.056
NAP	**5**	**250**	**-0.300**	**< 0.277**
Betula	6	300	-0.323	< 0.259
Chenopodiaceae	6	300	-0.041	< 0.889
Poaceae	6	300	-0.606	< 0.021
Artemesia	6	300	-0.475	< 0.086
Ligulaflorae	6	300	-0.537	< 0.048
NAP	**6**	**300**	**-0.305**	**< 0.288**

Table 8.17. Significant Bivariate Correlations between the Percentage of Plant Taxa and Contemporaneous Oxygen Isotope Values

Positive Correlations		
Pollen Type	Correlation (Pearson's *r*)	Significance (2-tailed)
Picea	+0.867	< 0.0001

Negative Correlations		
Pollen Type	Correlation (Pearson's *r*)	Significance (2-tailed)
Alnus	-0.581	< 0.007
Poaceae	-0.615	< 0.004
Artemesia	-0.588	< 0.006

Table 8.18. Insignificant Bivariate Correlations between the Percentage of Plant Taxa and Contemporaneous Oxygen Isotope Values

Positive Correlations		
Pollen Type	Correlation (Pearson's *r*)	Significance (2-tailed)
Pinus	+0.424	< 0.063
Chenopodiaceae	+0.003	< 0.989
AP	**+0.442**	**< 0.051**

Negative Correlations		
Pollen Type	Correlation (Pearson's *r*)	Significance (2-tailed)
Betula	-0.359	< 0.120
Ambrosia	-0.247	< 0.293
Aster	-0.257	< 0.274
Ligulaflorae	-0.314	< 0.177
NAP	**-0.442**	**< 0.051**

more Poaceae, *Alnus*, and *Artemesia* during times of greater relative moisture (when oxygen isotope values are relatively more negative), and more *Picea* during periods of aridity. Most taxa (i.e., *Pinus*, Chenopodiaceae, *Betula*, *Ambrosia*, *Aster,* and Ligulaflorae) show no significant relationship with contemporary climate conditions.

Again, climate is a variable that one would expect to be a leading variable, with effects that would not necessarily manifest themselves for years, decades, or centuries. Introducing lags into the bivariate models and calculating correlations produces interesting results, which are presented in Tables 8.19 and 8.20. Expectably, most of the taxa identified as significant in the coeval comparisons (*Alnus*, Poaceae, and *Artemesia,* but not *Picea*) remained significant over many of the short time intervals in the lagged model.

When vegetation is allowed to further lag climate, a few additional significant relationships emerge, as expected. *Pinus*, for instance, shows no significant relationship with the climate proxy until greater lags are introduced into the bivariate model. After 250 to 300 years, pine is

Table 8.19. Significant Bivariate Correlations between Oxygen Isotope (Leading) and Lagged Values of Taxonomic Abundance

Positive Correlations

Pollen Type	# Lags	Years	Correlation (Pearson's *r*)	Significance (2-tailed)
Picea	1	50	+0.910	< 0.0001
Picea	2	100	+0.811	< 0.0001
AP	**4**	**200**	**+0.658**	**< 0.006**
Pinus	5	250	+0.654	< 0.008
AP	**5**	**250**	**+0.713**	**< 0.003**
Pinus	6	300	+0.715	< 0.004
AP	**6**	**300**	**+0.710**	**< 0.004**

Negative correlations

Pollen Type	# Lags	Years	Correlation (Pearson's *r*)	Significance (2-tailed)
Poaceae	1	50	-0.587	< 0.008
Artemesia	1	50	-0.598	< 0.007
Poaceae	2	100	-0.599	< 0.009
Artemesia	3	150	-0.624	< 0.007
Aster	3	150	-0.708	< 0.001
Ambrosia	4	200	-0.683	< 0.004
Aster	4	200	-0.849	< 0.0001
NAP	**4**	**200**	**-0.658**	**< 0.006**
Chenopodiaceae	5	250	-0.775	< 0.001
Poaceae	5	250	-0.365	< 0.182
Ambrosia	5	250	-0.845	< 0.0001
Aster	5	250	-0.864	< 0.0001
NAP	**5**	**250**	**-0.713**	**< 0.003**
Chenopodiaceae	6	300	-0.859	< 0.0001
Ambrosia	6	300	-0.883	< 0.0001
Aster	6	300	-0.712	< 0.004
NAP	**6**	**300**	**-0.710**	**< 0.004**

**Table 8.20. Insignificant Bivariate Correlations between Oxygen Isotope (Leading)
and Lagged Values of Taxonomic Abundance**

Positive Correlations

Pollen Type	# Lags	Years	Correlation (Pearson's *r*)	Significance (2-tailed)
Pinus	1	50	+0.397	< 0.092
Chenopodiaceae	1	50	+0.029	< 0.906
AP	**1**	**50**	**+0.436**	**< 0.062**
Pinus	2	100	+0.431	< 0.074
AP	**2**	**100**	**+0.510**	**< 0.031**
Pinus	3	150	+0.491	< 0.045
Picea	3	150	+0.590	< 0.013
Ligulaflorae	3	150	+0.026	< 0.921
AP	**3**	**150**	**+0.580**	**< 0.015**
Pinus	4	200	+0.572	< 0.021
Picea	4	200	+0.264	< 0.322
Alnus	5	250	+0.005	< 0.986

Negative Correlations

Pollen Type	# Lags	Years	Correlation (Pearson's *r*)	Significance (2-tailed)
Betula	1	50	-0.363	< 0.127
Alnus	1	50	-0.533	< 0.019
Ambrosia	1	50	-0.262	< 0.278
Aster	1	50	-0.294	< 0.222
Ligulaflorae	1	50	-0.197	< 0.420
NAP	**1**	**50**	**-0.436**	**< 0.062**
Betula	2	100	-0.332	< 0.178
Alnus	2	100	-0.389	< 0.111
Chenopodiaceae	2	100	-0.090	< 0.722
Ambrosia	2	100	-0.367	< 0.134
Aster	2	100	-0.479	< 0.044
Ligulaflorae	2	100	-0.068	< 0.787
NAP	**2**	**100**	**-0.510**	**< 0.031**
Betula	3	150	-0.359	< 0.157
Alnus	3	150	-0.264	< 0.307
Chenopodiaceae	3	150	-0.289	< 0.260
Poaceae	3	150	-0.512	< 0.036
Ambrosia	3	150	-0.500	< 0.041
Polygonaceae	3	150	-0.112	< 0.668
NAP	**3**	**150**	**-0.580**	**< 0.015**

(table continued on next page)

**Table 8.20. Insignificant Bivariate Correlations between Oxygen Isotope (Leading)
and Lagged Values of Taxonomic Abundance (cont.)**

Negative Correlations

Pollen Type	# Lags	Years	Correlation (Pearson's *r*)	Significance (2-tailed)
Betula	4	200	-0.403	< 0.122
Alnus	4	200	-0.083	< 0.759
Chenopodiaceae	4	200	-0.579	< 0.019
Poaceae	4	200	-0.433	< 0.094
Artemesia	4	200	-0.560	< 0.024
Polygonaceae	4	200	-0.002	< 0.995
Ligulaflorae	4	200	-0.090	< 0.741
Betula	5	250	-0.492	< 0.063
Picea	5	250	-0.071	< 0.800
Poaceae	5	250	-0.365	< 0.182
Artemesia	5	250	-0.503	< 0.056
Ligulaflorae	5	250	-0.159	< 0.572
Betula	6	300	-0.657	< 0.011
Alnus	6	300	-0.053	< 0.856
Picea	6	300	-0.412	< 0.143
Poaceae	6	300	-0.258	< 0.374
Artemesia	6	300	-0.475	< 0.086
Aster	6	300	-0.712	< 0.004
Ligulaflorae	6	300	-0.330	< 0.249

significantly and positively correlated with oxygen isotopes, showing that a change to more arid climatic conditions promotes an expansion of pines on the landscape a few centuries later. This, of course, follows from the fact that pines require many years to both establish seedlings and have those individuals mature. This also makes sense since pines are a later-successional taxon that often takes time to pass through generations of lower-successional types before pine can establish itself.

Results in Table 8.19 show a relationship between negative oxygen isotope values (reflecting moist conditions), and future values of non-arboreal taxa including Poaceae, *Artemesia*, *Aster*, *Ambrosia*, and Chenopodiaceae, indicating that moist conditions promote the growth of these taxa. Additionally, a comparison of results from Tables 8.4 and 8.19 reveals that several statistically significant negative relationships between oxygen isotope values and plant taxa do not show up until lags are introduced into the model—three of the taxa listed above (*Aster*, *Ambrosia*, and Chenopodiaceae) had previously shown no significant contemporaneous relationship with the climate proxy.

Bivariate analyses had indicated that vegetation did respond to climate, but many types of plants remain unaffected until tens to hundreds of years have passed. What happens when all factors are considered together in

a multivariate procedure? To answer this, fire, climate, population, and vegetation proxies were submitted together in a time series analysis. Table 8.8 shows the resulting significant Granger causalities. Looking at this table, it is obvious that most of the unique significant predictors of future taxonomic abundances are the past values of other plant taxa. This occurs in 23 of the significant cases listed. In only 7 cases are non-vegetative factors identified as significant contributors in this multivariate analysis. For the most part, past vegetation is the primary factor controlling future vegetation.

Exogenous variables identified as significantly controlling future vegetation are: (1) charcoal, with a 50-year lead on Chenopodiaceae, (2) population with a 50-year lead on *Betula* and a 100-year lead on Chenopodiaceae, and (3) oxygen isotope values, with a 50-year lead on *Betula* and *Picea*, a 100-year lead on *Aster,* and both a 250- and 300-year lead on Chenopodiaceae. It is interesting to note that few taxa are identified by the algorithm as being influenced by fire or people. Climate influences a number of taxa, but is still not singled out as the only important factor determining plant types and their abundances on the landscape; historical constraints from past vegetation, fire, and population also shape plant communities.

Assessing Interactions: Comparing Bivariate and Multivariate Results

A repeating theme in these statistical results is that bivariate assessments of significance consistently identify more significant correlations between variables than do the multivariate Granger causality tests. This is because many of the factors that appear significant in the bivariate case become disqualified once other competing explanatory variables are simultaneously considered. It is the case that many of the variables in this analysis are intricately related to one another and multicollinearity runs through a large portion of the dataset.

Not only do many significant relationships exist between variables when lags are introduced into statistical models, but some significant relationships between variables only appear once lags are included. Historical constraints, then, are important. Taken together, these analyses show that the variables—taxonomic abundances, charcoal, population, and oxygen isotopes—all interact with one another on many different timescales. Variables are highly intertwined with one another, pointing to a system of feedbacks in which all components are linked together in complex ways. This demonstrates that humans, for example, are an integral part of the system, neither an external force imposing change on a "natural" system, nor passive victims of environmental change. Of all the variables considered, the climate proxy is the closest to being an external forcing mechanism on the system, as one would logically expect. Looking at Table 8.9, population and charcoal are listed as both leading and lagging variables; the climate proxy only shows up in a leading role (never as a result of the change in another variable).

Chapter 9. Conclusions

The goals of this volume have been to create a protocol, collect data, and examine the interaction between people, climate, and the landscape over a decadal to centennial time-scale. This time-scale was chosen for study because few research projects have dealt with interactions and change in landscapes on this temporal scale, although this is the scale at which succession, climate, and human impact primarily operate. Modern ecological studies examine interactions on annual to decadal-scales, and paleoecological studies mainly deal with processes on a millennial-scale, leaving a gap in our understanding of the important processes that operate across the intermediate time scales between these two major approaches.

This study focuses on the interactions of a natural and cultural system over the past thousand years in eastern Washington as one example for evaluating the processes interacting at decadal to centennial scales. In order to test and evaluate ideas about how people, vegetation, climate, and fire all affect each other, proxy variables had to be created to cover the necessary range of data. An existing population proxy, drawn from area archaeology, was used in this study. New variables covering taxonomic plant abundances over time (pollen), local climate (oxygen isotope records of aridity), and local incidence of vegetation fires (charcoal deposition) were generated, and matched to the resolution, scale, and extent of the pre-existing population proxy.

Using statistical analyses to explore the resulting dataset has yielded information on local environmental history, which is of use to people living in the area today, archaeologists, and paleoecologists who wish to understand both the historic and prehistoric record of human behavioral ecology in this part of the world. Additionally, statistical analyses have provided information on the general nature of landscape processes that operate on the decadal to centennial scale in this area. This is important for understanding past processes in other areas of our world and for understanding the probable changes and interactions that could take place in the next 50 to 250 years, given the events occurring now.

History of the Chief Joseph Dam Area of Eastern Washington

Area Vegetation History

This study has revealed several trends that are of regional or local interest and increased our knowledge of the ecological history of this particular area. Such paleovegetation data provide a finer scale than previous work, yielding a more suitable backdrop for understanding and evaluating human behavior. It was known, for instance, that this area had undergone changes in vegetation over the past few thousand years, but the details and timing of these vegetation changes were not as finely resolved as the archaeological record (Dalan 1985b).

The relatively high-resolution vegetation reconstruction generated by this study extends back 1,525 years, and is based on counts of pollen of different taxa recovered from samples of lake sediment from Five Lakes, a basin located approximately 13 kilometers north of Grand Coulee Dam and 6 kilometers southeast of the modern town of Nespelem. The Five Lakes core was sampled at 5 cm (approximately 53-year) intervals, and the resulting reconstruction of vegetation revealed a general trend over time towards decreased amounts of pine and increased amounts of early-seral taxa such as alder, birch, chenopods, and composites.

Previous pollen studies found that modern vegetation was established by 2,000 years ago (Chatters 1998) with a recent rise in Chenopodiaceae imposed upon this pollen zone about 100 years ago (Dalan 1985b, Davis et al. 1977). Examining the pollen diagram created for this study, the past 1,000 to 2,000 years of vegetation history can be described in finer terms than to simply call it "modern" in nature. The vegetation record can be further divided into two smaller pollen zones. The earliest subdivision from about 1,525 to 580 B.P. is a time of high and constant amounts of pine pollen, indicating that vegetation a thousand years ago was less diverse, that pine woodlands were more extensive than today, and that vegetation was more stable and unchanging than it is today.

Beginning about 580 B.P., vegetation became less stable, with individual taxa varying more in abundance from sample to sample. Overall, the recent portion of the vegetation record has also been dominated by pine, but at a lower and more fluctuating abundance than before. This upper, or more recent, part of the vegetation record is characterized by a larger contribution from non-arboreal pollen (herbs and shrubs) and a greater proportion of early-seral taxa as part of a more diverse and dynamic vegetation than before.

As predicted by other researchers, the recent portion of the record shows a growing abundance of Chenopodiceae. This spread of Chenopodiaceae was predicted based on the introduction of Euro-American land-use practices, primarily the introduction of large numbers of domestic grazing animals into eastern Washington starting around A.D. 1850 (Davis et al. 1977). Although a recent rise in chenopods is seen in this study, the magnitude of this rise is much lower than that documented in other records and the timing does not coincide with a date of A.D. 1850 but occurs much earlier in the Five Lakes pollen diagram—at approximately 580 B.P.

Area Fire History

Fire history in the area was investigated using the influx of charcoal particles into the same lake sediments used for pollen analysis. Using charcoal as a proxy for local vegetation fires, a 1,525-year history of area fire was reconstructed. Whether charcoal was measured in terms of the total influx of all sizes of particles, or simply the influx of large particles, results remained unchanged. The record shows that charcoal influx was greatest from 1,525 to 500 B.P., roughly the same time period as the early and stable pine-dominated vegetation described above.

If spikes in the influx of large charcoal particles (measured by #/cm^2/yr of charcoal particles 125-500 μm in diameter), represent individual fire events, then dates and return intervals between fire events can be determined. During the earlier part of the record (1,525 to 500 B.P.), fire occurred, on average, every 148 years and individual return intervals ranged from 94 to 232 years. These return rates are longer than those observed in modern ecological studies, because palynological charcoal records only record large scale events, while fire scars, soil profiles, and historic records are more sensitive and reflect information about stand-level processes. Results from Five Lakes do correspond well with return intervals that are recognizable in other, comparable, lacustrine charcoal records. Furthermore, comparing Five Lakes with other palynological studies, the length of the average return interval at Five Lakes (148 years) corresponds to returns seen for natural, rather than cultural, fire regimes (Chatters and Leavell 1994).

During the later part of this record (the past 500 years), charcoal influx was greatly reduced. This could signal a reduction in the size of fires in the area. Although a decrease in vegetation fires was expected during the era of active fire suppression in this area (from about 150 to 30 years ago [Martin et al. 1977]), it was a surprise to find that charcoal influx was low even before the advent of human fire suppression.

Area Climate History

As part of this study, oxygen isotope assays from lake carbonates were used to reconstruct local climate. The oxygen isotope record provided a reconstruction of relative aridity for the area, and could not be resolved into separate records for temperature and precipitation. The results from this analysis show that, overall, conditions were more arid in the study area at 1,000 B.P. than they are today. From 1,000 to about 600 B.P., conditions became progressively more moist (mesic) and reached a relative maximum about 600 B.P. This indicates that corresponding global period known as the Medieval Warm Period was expressed locally as an increase in precipitation, a decrease in evaporation, or both. After 600 B.P., local conditions became more arid. This drying episode, which peaked about 400 B.P., corresponds temporally with the global phenomenon known as the Little Ice Age. Results show that the LIA was expressed in the study area as a period of relative aridity rather than a period of cooler and moister conditions.

Processes Operating on Decadal to Centennial Scales

Interactions on the Landscape

One finding was that many of the statistically significant bivariate correlations between variables were shown to be unimportant once all variables were simultaneously considered in a multivariate procedure. Many variables that promised to have potential explanatory value when examined in isolation turned out to be linearly dependent on other factors when competing potential explanatory factors and time lags were entered into the analysis. This was one indication that human population, taxonomic abundances, fire, and climate participate in a highly complex set of interrelated feedback systems. In addition, many of the interrelationships were not apparent until lags of 50, 100, or even 300 years were entered into the model.

Once the dataset was evaluated using multivariate time series techniques (in the time domain), significant leading factors could be identified. Vegetation, for instance, was patterned mainly by past taxonomic abundances and oxygen isotope values. Of these two, past vegetation had a greater relative influence on future plant communities. History, then, constrains vegetation type more than fire, humans or climate—a surprising finding given that human activities and climate are thought of as primary determinants of change.

Although not in a primary role, population and fire did have some influence on future taxonomic abundances on short time-scales. The nature of the relationship between plants and humans was surprising, however. Times of high population were not, as expected, associated with relative increases in disturbance taxa. In fact, the opposite trend was observed, as human population sizes were significantly negatively correlated with disturbance taxa and positively correlated with later-successional taxa at several time scales.

Tests run on the charcoal proxy indicated that fire was also part of a densely interwoven pattern of relationships with population, vegetation, past fires, and climate. In multivariate models, fire was seen to be significantly patterned over the short term by human population sizes and over the long term by climate. Time series analysis and Granger causation tests indicated that, while population conditioned charcoal influx 50 years in the future, charcoal influxes likewise predicted population values 200 years in the future. Again, this demonstrates that the variables are complexly interrelated with one another, neither one clearly "causing" the other in a simplistic manner. Like vegetation, fire was also historically constrained by past values of itself (with significant autocorrelations).

By running and comparing bivariate and multivariate statistical analyses, significant interrelationships could be discovered and variables identified as leading or lagging in these interactions. Results indicated that no variables could be considered in isolation; human population, vegetation, and fire could not be considered extrinsic, independent factors imposing change on a system. Humans, for instance, influence fire, but fire regimes are seen to pattern population sizes in return. Of all the factors analyzed, oxygen isotope was identified most often as a linearly independent leading factor, showing that climate is the closest to being an "outside" (exogenous) factor in the past millennium in the study area. This analysis also showed the importance of historical constraints on this ecological system, which exert significant control on future conditions.

Appendix A. Abbreviations, Symbols, and Conventions Used

Table A.1. Abbreviations, Symbols, and Conventions Used

Notation	Meaning/Use
A.D.	Anno Domini, also know as the Common Era (C.E.) These are calendar years from historic records.
cSt	Centistokes
Local	Areas within 20 m of a basin (Jacobson and Bradshaw 1981)
Extralocal	Areas between 20 and several hundred m of a basin (roughly within a km)
Regional	Areas more than a few hundred m of a basin (usually several tens of km)
cal. A.D.	Calibrated radiocarbon dates, obtained by using dendro-calibration curves to convert radiocarbon dates to A.D. dates (using the CALIB. v.4.3 program, Dataset 2, after correcting for $\delta^{13}C$).
cal. B.C.	Before Christ, also know as Before the Common Era (B.C.E.) These are calibrated radiocarbon dates obtained by using dendro-calibration curves to convert radiocarbon dates to B.C. dates (using the CALIB v.4.3 program, Dataset 2, after correcting for $\delta^{13}C$).
AP	Arboreal Pollen—the sum of all tree pollen
NAP	Non-Arboreal Pollen—the sum of all herb and shrub (but not aquatic) pollen
LOI	Loss On Ignition analysis, in which the percent composition of organics and carbonates are determined via weight loss when samples are oxidized at two different temperatures in a muffle furnace (see Dean 1974 or Appendix C for procedures).
SEM	Scanning Electron Microscope The SEM and EDS used in this study were a JEOL 840A.
EDS	Energy Dispersive Spectrometry A detection system on an SEM which collects x-rays produced by materials when under the electron beam. X-ray counts for different energy levels are made, and x-ray "fingerprints" for different elements are used to identify the elemental composition of the material under the beam. This is similar to WDS, Wavelength Dispersive Spectrometry, a collector that counts the incidence of x-rays of different wavelengths.
AMS	Accelerator Mass Spectrometer AMS dates in this manuscript are those ^{14}C dates calculated from measurements of ^{14}C made using the accelerator at University of Arizona.
B.P.	Before Present—dates of B.P. reported here are measured in terms of uncalibrated ^{14}C years before present, where present is A.D. 1950, and ^{14}C dates have been corrected for $\delta^{13}C$ by the lab.
cm b.i.	centimeters below sediment-water interface (centimeters in depth in a lake core)
mm b.i.	millimeters below sediment-water interface (millimeters in depth in a lake core)
m asl	meters above sea level (modern elevation)

(continued on next page)

Table A.1. Abbreviations, Symbols, and Conventions Used (cont.)

Notation	Meaning/Use
μm	microns, micrometers
$\delta^{13}C$	carbon isotope value, reported in per mil (‰) $\delta^{13}C = 1000(R_{sample}/R_{standard} -1)$, where $R = {}^{13}C/{}^{12}C$ and the $R_{standard}$ is PDB-CO_2, with a $\delta^{13}C = 0$
$\delta^{18}O$	oxygen isotope value, reported in per mil (‰) $\delta^{18}O = 1000(R_{sample}/R_{standard} -1)$, where $R = {}^{18}O/{}^{16}O$ and the $R_{standard}$ is VSMOW-CO_2, with a $\delta^{18}O = 0$

Appendix B. Lake Locations

Table B.1. Location Information for Lakes Discussed in the Text

Lake Name	County in Washington State	Latitude	Longitude	UTM Principal Point 11	Elevation asl *	USGS 7.5' Map
Duley Lake	Okanogan	48°09'57"N	119°29'37"W	314594E 5337515N	736 m (2,414 ft)	JOE LAKE 1990
Five Lakes	Okanogan	48°04'55"N	118°55'46"W	356291E 5326990N	780 m (2,560 ft)	BELVEDERE 1990
Goose Lake	Okanogan	48°10'16"N	119°20'32"W	325852E 5337759N	372 m (1,220 ft)	BOOT MOUNTAIN 1980
Hidden Lake	Okanogan	48°08'10"N	119°20'03"W	326339E 5333858N	573 m (1,880 ft)	BOOT MOUNTAIN 1980
Rinker's Lake	Douglas	48°00'30"N	119°17'01"W	329672E 5319512N	704 m (2,310 ft)	TREFRY CANYON 1980

*asl = above modern sea level

Appendix C. Details of Laboratory Procedures

Table C.1. Loss-on-Ignition Analysis

Procedural Steps

Empty clean crucibles weighed
Crucibles filled with approximately 2 cm^3 wet sediment
Crucibles and wet sediment placed in 80 $^{\circ}$C drying oven for 12 hours
Crucibles and dry sediment allowed to cool to room temperature
Crucibles and dry sediment weighed to 0.0001 g
Crucibles and dry sediment put in 550 $^{\circ}$C muffle furnace for 90 minutes
Crucibles and (post 550 $^{\circ}$C burn) sediment allowed to cool to room temperature
Crucibles and (post 550 $^{\circ}$C burn) sediment weighed to 0.0001 g
Crucibles and (post 550 $^{\circ}$C burn) sediment put in 1,000 $^{\circ}$C muffle furnace for 90 minutes
Crucibles and (post 1,000 $^{\circ}$C burn) sediment allowed to cool to room temperature
Crucibles and (post 1,000 $^{\circ}$C burn) sediment weighed to 0.0001 g
Residue scanned under 40x binocular dissecting scope for inorganic sediment identification
Residue put in labeled coin envelopes for storage and crucibles cleaned

---

Table C.2 Pollen Processing

Procedural Steps

One cm^3 of sediment was taken from a 1-cm span of stratigraphic depth within the core
Sediment placed in a 50 ml Nalgene centrifuge tube
2 *Lycopodium* aliquot tablets added to tube (lot #124961)*
3 ml 10% HCl added to dissolved aliquot tablets — allowed to stand overnight
50 ml 10% HCl added, centrifuged, decanted
50 ml concentrated HCl in hot** water bath for 10 minutes, centrifuged, decanted
water wash—50 ml distilled water, centrifuged, decanted
50 ml 10% KOH in hot water bath for 10 minutes, centrifuged, decanted
50 ml water, centrifuged, decanted
    (repeated about six times until decant was clear and pH neutral)
50 ml 10% HCl, centrifuged, decanted
sieved (through sterilized metal window screen or fine tulle to remove large particles)
filtered through 8 μm nylon mesh (a.k.a hematology screen or "Nytex") using a motorized vacuum pump
filtrate discarded and 8 μm and larger particles washed into test tubes with distilled water
25 ml concentrated HF in hot water bath for 30 minutes, centrifuged, decanted
50 ml distilled water, centrifuged, decanted (twice)
glycerin preparation slides made from samples to assess processing efficacy and possible damage to pollen
optional step for samples with sulfides: 50 ml H_2NO_3, centrifuged and decanted followed by
    a distilled water wash
optional step for flocculated samples: 50 ml 5% sodium hexametaphosphate and 10 seconds of
    sonification, centrifuged, decanted, followed by a distilled water wash
optional step for siliceous samples: 50 ml HF in hot water bath for 30 minutes, centrifuged, decanted,
    followed by two distilled water washes
50 ml concentrated glacial acetic acid, centrifuged and decanted (twice)
25 ml acetolysis mixture, in hot water bath for 7 minutes
50 ml concentrated glacial acetic acid, centrifuged and decanted (twice)
50 ml distilled water, centrifuged, decanted
preparation (smear) slides of all samples made with glycerin mount—evaluated under microscope
50 ml ethanol, centrifuged, decanted
50 ml tert-butanol (TBA) centrifuged, decanted
samples pipetted into vials with silicon oil (10,000 cSt) and TBA allowed to evaporate

---

*for batch #124961 *Lycopodium* spore tables, the average number of spores per tablet was 12,542
**hot = 80 °C

Appendix D. LOI Results

Table D.1. LOI Results for Five Lakes

Depth (cm b.i.)	Weight of Crucible (g)	Weight of Crucible & Dry Sediment (g)	Weight of Crucible & Sediment after 550 °C Burn (g)	Weight of Crucible & Sediment after 1,000 °C Burn (g)	% Organics	% Carbonates	% Other*
10	8.5031	8.9385	8.8431	8.8141	21.91	15.14	62.95
11	8.5054	8.8062	8.7470	8.7274	19.68	14.81	65.51
12	8.8678	9.1431	9.0859	9.0662	20.78	16.26	62.96
13	8.2245	8.4663	8.4160	8.3970	20.80	17.86	61.34
14	8.3830	8.5860	8.5413	8.5273	22.02	15.67	62.31
15	8.0094	8.2064	8.1640	8.1509	21.52	15.11	63.36
16	8.3680	8.5673	8.5263	8.5142	20.57	13.80	65.63
17	8.5178	8.9810	8.8842	8.8578	20.90	12.95	66.15
18	8.5478	8.8172	8.7673	8.7493	18.52	15.19	66.29
19	8.5100	8.8034	8.7516	8.7321	17.66	15.11	67.24
20	8.3596	8.6951	8.6381	8.6156	16.99	15.24	67.77
21	8.4644	8.7488	8.6994	8.6821	17.37	13.82	68.81
22	9.2716	9.5955	9.5390	9.5196	17.44	13.61	68.94
23	8.2078	8.6121	8.5427	8.5187	17.17	13.49	69.34
24	8.4810	8.8401	8.7787	8.7575	17.10	13.42	69.48
25	8.1791	8.5863	8.5253	8.4934	14.98	17.80	67.22
26	7.7511	8.1450	8.0840	8.0566	15.49	15.81	68.70
27	8.5166	8.9171	8.8495	8.8251	16.88	13.85	69.27
28	8.6897	9.2391	9.1492	9.1163	16.36	13.61	70.03
29	8.4757	8.8873	8.8157	8.7955	17.40	11.15	71.45
30	8.3056	8.7422	8.6732	8.6501	15.80	12.02	72.18
31	8.3860	8.6786	8.6359	8.6160	14.59	15.46	69.95
32	9.0077	9.4851	9.4158	9.3850	14.52	14.66	70.82
33	8.5699	9.0745	8.9996	8.9702	14.84	13.24	71.91
34	8.9870	9.4852	9.4119	9.3838	14.71	12.82	72.47
35	8.2281	8.6441	8.5788	8.5573	15.70	11.75	72.56
36	8.4250	8.7992	8.7430	8.7218	15.02	12.88	72.11
37	8.6041	9.0290	8.9643	8.9418	15.23	12.03	72.74
38	8.3570	8.8687	8.7932	8.7666	14.75	11.81	73.44
39	8.2296	8.7446	8.6654	8.6424	15.38	10.15	74.47
40	7.8852	8.4634	8.3675	8.3405	16.59	10.61	72.80
41	8.3164	8.7421	8.6620	8.6374	18.82	13.13	68.05
42	8.2363	8.7034	8.6041	8.5766	21.26	13.38	65.36
43	8.2378	8.5507	8.4707	8.4564	25.57	10.39	64.05
44	8.5913	8.8481	8.7823	8.7663	25.62	14.16	60.22
45	8.3811	8.5697	8.5075	8.4965	32.98	13.26	53.76
46	7.9573	8.2128	8.1094	8.0936	40.47	14.05	45.48
47	8.4481	8.6300	8.5480	8.5372	45.08	13.49	41.43
48	8.4132	8.5277	8.4693	8.4622	51.00	14.09	34.90
49	8.2252	8.3753	8.2961	8.2879	52.76	12.42	34.82

(continued on next page)

Table D.1. LOI Results for Five Lakes (cont.)

Depth (cm b.i.)	Weight of Crucible (g)	Weight of Crucible & Dry Sediment (g)	Weight of Crucible & Sediment after 550 °C Burn (g)	Weight of Crucible & Sediment after 1,000 °C Burn (g)	% Organics	% Carbonates	% Other*
50	8.3835	8.5254	8.4531	8.4452	50.95	12.65	36.40
51	8.0102	8.1467	8.0817	8.0738	47.62	13.15	39.23
52	8.3693	8.4921	8.4332	8.4264	47.96	12.59	39.45
53	8.5194	8.6791	8.5975	8.5900	51.10	10.67	38.23
54	8.5506	8.7066	8.6262	8.6186	51.54	11.07	37.39
55	8.5114	8.6882	8.6085	8.5924	45.08	20.70	34.22
56	8.3603	8.5899	8.4766	8.4666	49.35	9.90	40.75
57	8.4671	8.6053	8.5431	8.5332	45.01	16.28	38.71
58	9.2730	9.4759	9.3790	9.3678	47.76	12.55	39.70
59	8.2107	8.3922	8.3064	8.2969	47.27	11.90	40.83
60	8.4825	8.7194	8.6039	8.5920	48.75	11.42	39.83
61	8.1837	8.400	8.3072	8.2862	42.90	22.07	35.03
62	7.7536	7.9173	7.8400	7.8272	47.22	17.77	35.01
63	8.5201	8.6771	8.5971	8.5894	50.96	11.15	37.90
64	8.6933	8.8641	8.7837	8.7760	47.07	10.25	42.68
65	8.4800	8.6756	8.6043	8.5904	36.45	16.15	47.40
66	8.3112	8.6323	8.5220	8.4956	34.35	18.69	46.96
67	8.3897	8.5828	8.4983	8.4810	43.76	20.36	35.88
68	9.0109	9.2171	9.1247	9.1084	44.81	17.97	37.22
69	8.5738	8.7460	8.6676	8.6547	45.53	17.03	37.45
70	8.9893	9.1950	9.0965	9.0829	47.89	15.03	37.08
71	8.2358	8.4613	8.3610	8.3448	44.48	16.33	39.19
72	8.4261	8.6660	8.5744	8.5535	38.18	19.80	42.02
73	8.3647	8.7057	8.5914	8.5528	33.52	25.73	40.75
74	8.4675	8.7014	8.6178	8.5947	35.74	22.45	41.81
75	8.5270	8.8018	8.7054	8.6808	35.08	20.35	44.57
76	8.2440	8.7043	8.5352	8.4974	36.74	18.66	44.60
77	8.3909	8.6824	8.5632	8.5413	40.89	17.07	42.04
78	8.3753	8.5913	8.5016	8.4845	41.53	17.99	40.48
79	8.2922	8.5130	8.4228	8.4010	40.85	22.44	36.71
80	8.3176	8.5247	8.4335	8.4144	44.04	20.96	35.00
81	8.6717	8.9390	8.8559	8.8301	31.09	21.94	46.97
82	8.5036	8.7975	8.7084	8.6842	30.32	18.71	50.97
83	8.5060	8.8931	8.7812	8.7474	28.91	19.84	51.25
84	8.8685	9.4120	9.2627	9.2145	27.47	20.16	52.37
85	8.4140	8.7764	8.6801	8.6438	26.57	22.76	50.66
86	8.4493	8.7626	8.6786	8.6504	26.81	20.46	52.73
87	7.9580	8.3984	8.2814	8.2428	26.57	19.92	53.51
88	8.3817	8.7976	8.6875	8.6500	26.47	20.49	53.04
89	8.5925	8.9956	8.8864	8.8496	27.09	20.75	52.16
90	8.2400	8.7712	8.6294	8.5776	26.69	22.16	51.14
91	8.2378	8.6623	8.5531	8.5017	25.72	27.52	46.76

(continued on next page)

Table D.1. LOI Results for Five Lakes (cont.)

Depth (cm b.i.)	Weight of Crucible (g)	Weight of Crucible & Dry Sediment (g)	Weight of Crucible & Sediment after 550 °C Burn (g)	Weight of Crucible & Sediment after 1,000 °C Burn (g)	% Organics	% Carbonates	% Other*
92	8.3180	8.6269	8.5360	8.5065	29.43	21.70	48.87
93	7.8874	8.1831	8.0972	8.0715	29.05	19.75	51.20
94	8.2310	8.5692	8.4801	8.4490	26.35	20.90	52.76
95	8.3600	8.6988	8.6062	8.5769	27.33	19.65	53.01
96	8.6060	8.9354	8.8482	8.8193	26.47	19.94	53.59
97	8.2261	8.6374	8.5318	8.4892	25.67	23.54	50.79
98	8.3846	8.6987	8.6121	8.5847	27.57	19.83	52.60
99	8.0117	8.2051	8.1483	8.1336	29.37	17.27	53.36
100	8.3703	8.6050	8.5367	8.5185	29.10	17.62	53.27
101	8.5205	8.7917	8.7134	8.6918	28.87	18.10	53.03
102	8.5516	8.8443	8.7620	8.7345	28.12	21.35	50.53
103	8.5137	8.7597	8.6933	8.6653	26.99	25.87	47.14
104	8.3628	8.5740	8.5162	8.4923	27.37	25.72	46.91
105	8.4674	8.6574	8.6015	8.5801	29.42	25.60	44.98
106	9.2741	9.4158	9.3546	9.3435	43.19	17.80	39.01
107	8.2104	8.3530	8.2890	8.2786	44.88	16.58	38.54
108	8.4836	8.6431	8.5770	8.5606	41.44	23.37	35.19
109	8.1841	8.2930	8.2512	8.2420	38.38	19.20	42.42
110	7.7540	7.9714	7.8922	7.8775	36.43	15.37	48.20
111	8.5205	8.7692	8.6920	8.6780	31.04	12.79	56.16
112	8.6940	8.9785	8.8877	8.8708	31.92	13.50	54.58
113	8.4797	8.6709	8.6026	8.5874	35.72	18.07	46.21
114	8.3118	8.4832	8.4182	8.4044	37.92	18.30	43.78
115	8.3900	8.6373	8.5544	8.5238	33.52	28.12	38.36
116	9.0127	9.2211	9.1526	9.1300	32.87	24.65	42.48
117	8.5750	8.7896	8.7127	8.6925	35.83	21.39	42.77
118	8.9902	9.2058	9.1261	9.1071	36.97	20.03	43.00
119	8.2365	8.4583	8.3759	8.3559	37.15	20.49	42.36
120	8.4275	8.7686	8.6364	8.6063	38.76	20.06	41.19
121	8.3664	8.6286	8.5223	8.4945	40.54	24.10	35.36
122	8.4695	8.7740	8.6572	8.6292	38.36	20.90	40.74
123	8.5283	8.7600	8.6743	8.6558	36.99	18.15	44.87
124	8.2454	8.6207	8.5070	8.4789	30.30	17.02	52.69
125	8.3927	8.7222	8.6262	8.6033	29.14	15.80	55.07
126	8.3765	8.7524	8.6464	8.6189	28.20	16.63	55.17
127	8.2930	8.5872	8.5070	8.4759	27.26	24.03	48.71
128	8.3184	8.6832	8.5787	8.5515	28.65	16.95	54.41
129	8.6718	9.0672	8.9528	8.9242	28.93	16.44	54.63
130	8.5045	8.8551	8.7498	8.7244	30.03	16.47	53.50
131	8.5073	8.7952	8.7059	8.6860	31.02	15.71	53.27
132	8.8706	9.1571	9.0686	9.0480	30.89	16.34	52.77
133	8.4152	8.8119	8.7036	8.6676	27.30	20.62	52.08

(continued on next page)

Table D.1. LOI Results for Five Lakes (cont.)

Depth (cm b.i.)	Weight of Crucible (g)	Weight of Crucible & Dry Sediment (g)	Weight of Crucible & Sediment after 550 °C Burn (g)	Weight of Crucible & Sediment after 1,000 °C Burn (g)	% Organics	% Carbonates	% Other*
134	8.4500	8.7516	8.6668	8.6405	28.12	19.82	52.06
135	7.9590	8.3356	8.2330	8.1998	27.24	20.04	52.72
136	8.3829	8.7017	8.6155	8.5877	27.04	19.82	53.14
137	8.5938	8.8799	8.7914	8.7625	30.93	22.96	46.11
138	8.2413	8.6498	8.5210	8.4739	31.53	26.20	42.27
139	8.2390	8.6135	8.4950	8.4518	31.64	26.22	42.14
140	8.3189	8.6196	8.5224	8.4916	32.32	23.28	44.40
141	7.8887	8.2370	8.1289	8.0944	31.04	22.51	46.45
142	8.2743	8.5305	8.4459	8.4300	33.05	14.11	52.85
143	8.3603	8.7167	8.6085	8.5633	30.36	28.82	40.82
144	8.6074	8.9652	8.8543	8.8094	30.99	28.52	40.48
145	8.2835	8.8212	8.6503	8.5813	31.78	29.16	39.05

*% Other = 100 - %organics - %carbonates = the index of terrigenous input

Table D.2. LOI Results for Hidden Lake

Depth (cm b.i.)	Weight of Crucible (g)	Weight of Crucible & Dry Sediment (g)	Weight of Crucible & Sediment after 550 °C Burn (g)	Weight of Crucible & Sediment after 1,000 °C Burn (g)	% Organics	% Carbonates	% Other*
0.5	8.1061	8.3113	8.2612	8.2447	24.42	18.27	57.31
1.5	8.8621	9.2799	9.2072	9.1828	17.40	13.27	69.33
2.5	8.4087	8.9551	8.8807	8.8584	13.62	9.28	77.11
3.5	8.7523	9.4400	9.3549	9.3266	12.37	9.35	78.27
4.5	8.2847	8.8583	8.7946	8.7709	11.11	9.39	79.50
5.5	8.3102	9.3628	9.2704	9.2449	8.78	5.51	85.72
6.5	8.4755	8.7668	8.6477	8.6304	40.89	13.50	45.62
7.5	8.2332	8.4871	8.3960	8.3824	35.88	12.17	51.95
8.5	8.3007	8.5836	8.4616	8.4477	43.12	11.17	45.71
9.6	8.3774	8.7611	8.6269	8.6050	34.98	12.97	52.05
10.5	8.0043	8.7675	8.6893	8.6670	10.25	6.64	83.11
11.5	8.2020	8.7100	8.6503	8.6316	11.75	8.37	79.88
12.5	8.2240	8.4982	8.3971	8.3757	36.87	17.74	45.39
13.5	8.5000	8.9524	8.8297	8.7935	27.12	18.19	54.69
14.5	8.6614	9.1866	9.1120	9.0765	14.20	15.36	70.43
25.0	8.0445	8.6108	8.5435	8.4574	11.88	34.55	53.56
50.0	7.8560	8.8880	8.7897	8.6351	9.53	34.05	56.43
56.0	8.4111	9.4906	9.3850	9.2308	9.78	32.46	57.75

*% Other = 100 - %organics - %carbonates = the index of terrigenous input

Table D.3. LOI Results for Rinker's Lake

Depth (cm b.i.)	Weight of Crucible (g)	Weight of Crucible & Dry Sediment (g)	Weight of Crucible & Sediment after 550 °C Burn (g)	Weight of Crucible & Sediment after 1,000 °C Burn (g)	% Organics	% Carbonates	% Other*
55	9.2222	11.1241	11.0246	10.827	5.23	23.61	71.15

*% Other = 100 - %organics - %carbonates = the index of terrigenous input

Appendix E. Oxygen Isotope Results

**Table E.1. Isotope Sample Description and Results
from Hidden and Rinker's Lakes**

Lake	Depth (cm b.i.)	Age B.P.	Dry weight (g)	$\delta^{13}C$ Reported by Lab (‰)	$\delta^{18}O$ Reported by Lab (‰)
Hidden	0	0	0.2	-0.98	-5.00
Hidden	5	100	1.0	-1.08	-4.70
Hidden	10	200	0.5	-0.82	-4.69
Hidden	15	300	0.8	-0.91	-4.92
Hidden	20	400	0.6	-0.93	-4.59
Hidden	25	500	0.7	-0.82	-4.65
Hidden	30	600	0.7	-0.76	-4.96
Hidden	35	700	0.8	-0.90	-4.52
Hidden	40	800	0.5	-0.65	-3.90
Hidden	45	900	0.6	-0.92	-4.04
Hidden	50	1,000	1.7	-0.75	-3.94
Rinker's	0	0	0.4	-1.53	-9.87
Rinker's	5	157	0.6	-1.41	-8.47
Rinker's	10	313	0.5	-1.26	-6.46
Rinker's	15	470	0.5	-1.51	-6.71
Rinker's	20	626	0.7	-1.01	-8.61
Rinker's	25	783	0.9	-1.03	-8.24
Rinker's	30	939	1.1	-0.57	-6.79
Rinker's	35	1,096	1.0	-0.68	-6.75
Rinker's	40	1,252	1.2	-0.96	-7.47
Rinker's	45	1,409	1.6	-1.06	-7.06
Rinker's	50	1,570	1.1	-1.37	-7.13

Table E.2. Interpolated Oxygen Isotope Values for each 50-year Study Interval
(from Rinker's and Hidden Lakes)

Date cal A.D.	Hidden Lake Interpolated $\delta^{18}O$	Rinker's Lake Interpolated $\delta^{18}O$	Z-score for Hidden Lake Isotope Values	Z-score for Rinker's Lake Isotope Values	Composite Isotope Index*
1900	-4.70	-8.98	-0.58	-1.88	-1.23
1850	-4.70	-8.45	-0.58	-1.20	-0.89
1800	-4.70	-7.92	-0.58	-0.51	-0.55
1750	-4.80	-7.28	-0.84	0.33	-0.25
1700	-4.90	-6.63	-1.09	1.17	0.04
1650	-4.75	-6.62	-0.71	1.19	0.24
1600	-4.60	-6.60	-0.33	1.21	0.44
1550	-4.65	-6.84	-0.46	0.89	0.22
1500	-4.70	-7.08	-0.58	0.58	0.00
1450	-4.85	-7.69	-0.96	-0.20	-0.58
1400	-5.00	-8.29	-1.34	-0.99	-1.16
1350	-4.75	-8.37	-0.71	-1.08	-0.90
1300	-4.50	-8.44	-0.08	-1.18	-0.63
1250	-4.20	-8.26	0.67	-0.95	-0.14
1200	-3.90	-8.08	1.43	-0.71	0.36
1150	-3.95	-7.62	1.30	-0.11	0.59
1100	-4.00	-7.15	1.18	0.49	0.83
1050	-3.95	-6.96	1.30	0.74	1.02
1000	-3.90	-6.77	1.43	0.99	1.21
950	-3.85	-6.58	1.55	1.23	1.39
	$\bar{x} = -4.47$	$\bar{x} = -7.53$			
	$s_x = 0.40$	$s_x = 0.77$			

Appendix F. Pollen Data from Five Lakes

Table F.1. Raw Counts of Arboreal Pollen Grains from Five Lakes

Depth (cm b.i.)	*Lycopodium* aliquot	*Pinus* undiff.	*Pinus* haploxylon	*Pinus* diploxylon	*Pinus* half* undiff.	*Larix/ Pseudotsuga*
0	1,220	175	0	11	223	0
5	316	173	6	67	186	0
10	160	106	20	37	103	2
15	475	238	30	35	229	0
20	135	116	44	51	70	2
25	524	147	43	28	126	0
30	189	149	37	71	115	6
35	316	314	11	40	120	2
40	223	273	3	30	129	2
45	498	314	3	36	157	0
50	1,238	271	3	43	186	0
55	884	268	2	19	59	0
60	728	518	0	21	206	0
65	749	312	7	35	256	2
70	420	421	3	54	88	0
75	413	226	0	9	259	0
80	213	248	3	52	195	0
85	157	250	7	47	219	0
90	234	333	1	44	223	2
95	119	192	6	51	181	0
100	187	388	1	36	276	0
105	516	315	4	54	202	0
110	439	306	2	81	150	0
115	333	238	3	36	207	0
120	569	325	0	20	239	0
125	235	240	3	38	208	2
130	440	281	0	8	288	0
135	272	292	0	13	297	0
140	255	297	1	44	176	0
145	369	301	0	19	239	0

*Note: grains counted are whole grains except where labeled as "half"

Table F.1. Raw Counts of Arboreal Pollen Grains from Five Lakes (cont.)

Depth (cm b.i.)	Acer	Betula	Alnus	Tsuga	Abies	Abies half*	Picea	Picea half*	Quercus
0	0	13	10	10	1	0	4	1	7
5	0	34	26	6	0	0	3	6	5
10	2	39	30	2	4	1	7	4	5
15	2	18	26	6	5	0	8	4	1
20	2	26	21	12	4	0	4	1	8
25	1	36	15	8	3	0	7	0	7
30	8	18	12	4	2	0	9	2	6
35	6	24	20	4	1	0	14	0	5
40	3	11	6	4	2	0	6	3	3
45	1	19	19	6	0	0	5	3	1
50	2	6	14	4	0	0	3	2	1
55	1	3	14	9	4	0	5	1	0
60	1	3	13	7	1	0	7	1	0
65	0	6	15	10	0	0	8	5	1
70	0	7	18	6	7	0	10	0	0
75	1	3	3	5	1	0	8	8	0
80	0	1	10	8	0	0	9	4	1
85	0	1	3	7	1	0	12	1	1
90	0	7	18	10	1	0	13	1	1
95	0	4	10	3	0	0	8	6	1
100	1	15	19	6	1	0	15	4	0
105	0	8	16	12	1	0	8	1	2
110	1	7	13	9	2	0	12	2	1
115	2	15	15	11	0	0	14	9	0
120	3	10	11	9	0	0	11	11	1
125	2	11	12	9	6	1	9	8	2
130	1	5	17	9	0	0	7	3	2
135	1	5	11	8	0	0	14	2	2
140	0	1	8	11	4	1	9	5	2
145	1	2	5	9	1	0	6	1	0

*Note: grains counted are whole grains except where labeled as "half"

Table F.2. Raw Counts of Non-arboreal Pollen Grains from Five Lakes

Depth (cm b.i.)	Poaceae	*Artemesia-*type	*Ambrosia-*type	*Aster-*type	Chenopo-diaceae	Liguliflorae-type	Polygon-aceae	Apiaceae
0	15	14	11	8	23	0	2	2
5	38	78	6	15	25	0	1	0
10	65	73	10	12	32	1	4	2
15	38	59	9	15	25	3	0	0
20	36	35	13	8	29	0	1	0
25	28	47	8	8	27	1	1	1
30	46	37	18	19	46	1	0	3
35	44	53	13	25	48	0	0	0
40	26	56	9	22	7	0	0	3
45	35	31	7	5	12	1	0	0
50	31	30	2	3	5	0	0	2
55	47	32	1	4	4	0	1	0
60	18	10	9	1	2	0	0	0
65	30	15	0	3	7	0	0	0
70	24	26	2	7	5	1	0	0
75	10	8	1	0	1	0	0	0
80	15	11	0	2	6	0	0	1
85	12	10	0	0	11	1	0	0
90	28	8	2	1	15	0	0	0
95	26	11	2	0	16	0	0	0
100	33	18	4	0	20	0	0	0
105	28	14	0	4	5	1	0	0
110	21	21	0	3	32	0	0	0
115	41	18	1	0	7	0	0	0
120	25	28	3	2	5	1	0	0
125	27	19	3	6	7	0	0	0
130	14	11	3	3	8	0	0	0
135	26	17	1	1	5	0	0	0
140	24	10	0	0	6	0	0	0
145	5	8	2	2	7	0	0	0

Table F.3. Raw Counts of Unidentified/Unidentifiable Pollen Grains (not included in pollen sums) from Five Lakes

Depth (cm b.i.)	Crumpled	Hidden	Corroded	Degraded	Broken	Unknown
0	4	2	5	23	1	8
5	1	1	0	18	0	2
10	5	3	2	18	0	13
15	1	1	1	12	0	7
20	3	4	2	12	1	7
25	7	2	1	19	0	3
30	5	1	2	31	0	22
35	2	0	0	30	0	8
40	5	0	0	22	0	5
45	2	1	0	14	0	15
50	0	1	0	11	0	2
55	2	0	0	9	0	10
60	0	0	0	1	0	2
65	0	0	1	5	0	2
70	0	0	1	2	0	13
75	0	0	0	3	0	2
80	0	1	1	4	0	2
85	1	0	0	0	0	2
90	0	0	0	0	0	3
95	0	0	0	3	0	3
100	0	0	0	1	0	4
105	0	2	0	1	0	5
110	0	1	0	1	0	3
115	0	0	0	2	0	4
120	0	0	0	7	0	1
125	0	0	0	8	0	2
130	0	0	0	1	0	7
135	0	0	0	1	0	0
140	0	0	0	0	0	5
145	0	0	0	0	0	0

Table F.4. Raw Counts of Aquatics (not included in pollen sums) from Five Lakes

Depth (cm b.i.)	*Lycopodium* aliquot	years BP uncalibrated ^{14}C yrs	*Carex*	*Typha*	*Pediastrum*	*Ruppia*	Cupressaceae
0	1,220	0	11	0	0	1	0
5	316	58	6	0	4	5	0
10	233	116	17	0	1	3	2
15	475	174	9	10	0	3	2
20	135	232	5	0	1	7	0
25	524	290	16	0	1	4	0
30	189	348	14	0	0	0	0
35	316	406	10	6	1	2	0
40	223	464	8	0	2	0	0
45	498	522	23	1	2	0	0
50	1,238	580	17	0	1	0	0
55	884	628	18	0	0	0	0
60	420	676	21	12	26	0	0
65	749	724	14	4	26	2	0
70	728	772	1	3	5	0	0
75	413	820	4	5	30	0	0
80	147	866	3	0	15	0	0
85	157	912	18	4	47	0	0
90	234	958	9	0	163	1	0
95	119	1,004	26	0	79	0	0
100	187	1,050	26	5	75	0	0
105	516	1,105	12	4	68	0	0
110	439	1,160	21	20	850	0	0
115	333	1,215	10	0	62	5	0
120	569	1,270	13	0	94	1	0
125	235	1,325	25	4	185	2	0
130	440	1,378	9	5	86	5	0
135	272	1,431	4	4	68	4	0
140	255	1,484	29	10	719	1	0
145	369	1,537	1	4	166	0	0

Table F.5. Percent of Arboreal Taxa in Samples from Five Lakes

Depth (cm b.i.)	Pinus	Larix/ Pseudo-tsuga	Acer	Betula	Alnus	Tsuga	Abies	Picea	Quercus
0	71.17	0.00	0.00	3.11	2.39	2.39	0.24	1.08	1.67
5	53.80	0.00	0.00	6.54	5.00	1.15	0.00	1.15	0.96
10	42.48	0.40	0.40	7.72	5.94	0.40	0.89	1.39	0.99
15	65.80	0.00	0.32	2.84	4.10	0.95	0.79	1.58	0.16
20	54.97	0.45	0.45	5.81	4.69	2.68	0.89	1.01	1.79
25	58.66	0.00	0.21	7.52	3.13	1.67	0.63	1.46	1.46
30	57.17	1.09	1.45	3.27	2.18	0.73	0.36	1.81	1.09
35	62.13	0.29	0.88	3.51	2.92	0.58	0.15	2.05	0.73
40	69.64	0.38	0.56	2.07	1.13	0.75	0.38	1.41	0.56
45	75.04	0.00	0.17	3.30	3.30	1.04	0.00	1.13	0.17
50	79.77	0.00	0.39	1.17	2.72	0.78	0.00	0.78	0.19
55	71.73	0.00	0.23	0.68	3.15	2.03	0.90	1.24	0.00
60	89.85	0.00	0.14	0.42	1.82	0.98	0.14	1.05	0.00
65	82.89	0.34	0.00	1.03	2.58	1.72	0.00	1.81	0.17
70	82.20	0.00	0.00	1.10	2.83	0.94	1.10	1.57	0.00
75	89.01	0.00	0.24	0.73	0.73	1.22	0.24	2.93	0.00
80	85.85	0.00	0.00	0.21	2.14	1.71	0.00	2.36	0.21
85	87.34	0.00	0.00	0.21	0.63	1.48	0.21	2.74	0.21
90	82.13	0.34	0.00	1.17	3.02	1.68	0.17	2.27	0.17
95	80.19	0.00	0.00	0.94	2.36	0.71	0.00	2.59	0.24
100	80.77	0.00	0.14	2.15	2.73	0.86	0.14	2.44	0.00
105	82.65	0.00	0.00	1.39	2.79	2.09	0.17	1.48	0.35
110	79.11	0.00	0.17	1.19	2.22	1.53	0.34	2.13	0.17
115	74.75	0.00	0.39	2.95	2.95	2.16	0.00	3.63	0.00
120	80.22	0.00	0.52	1.73	1.90	1.55	0.00	2.85	0.17
125	76.24	0.40	0.40	2.18	2.38	1.78	1.39	2.57	0.40
130	84.16	0.00	0.19	0.97	3.30	1.75	0.00	1.65	0.39
135	83.29	0.00	0.18	0.92	2.02	1.47	0.00	2.57	0.37
140	84.65	0.00	0.00	0.20	1.57	2.17	0.89	2.26	0.39
145	90.06	0.00	0.20	0.41	1.02	1.84	0.20	1.33	0.00

Table F.6. Percent of Non-arboreal Taxa in Samples from Five Lakes

Depth (cm b.i.)	Cheno-podiaceae	Poaceae	*Artemesia*-type	*Ambrosia*-type	Polygon-aceae	Apiaceae	Other Composites (Asteraceae*)	Pollen Sum**
0	5.50	3.59	3.35	2.63	0.48	0.48	4.55	4.55
5	4.81	7.31	15.01	1.15	0.19	0.00	4.04	4.04
10	6.34	12.87	14.46	1.98	0.79	0.40	4.55	4.55
15	3.94	5.99	9.30	1.42	0.00	0.00	4.26	4.26
20	6.48	8.04	7.82	2.91	0.22	0.00	4.69	4.69
25	5.64	5.85	9.81	1.67	0.21	0.21	3.55	3.55
30	8.35	8.35	6.72	3.27	0.00	0.54	6.90	6.90
35	7.02	6.43	7.75	1.90	0.00	0.00	5.56	5.56
40	1.32	4.89	10.53	1.69	0.00	0.56	5.83	5.83
45	2.09	6.09	5.39	1.22	0.00	0.00	2.26	2.26
50	0.97	6.03	5.84	0.39	0.00	0.39	0.97	0.97
55	0.90	10.59	7.21	0.23	0.23	0.00	1.13	1.13
60	0.28	2.52	1.40	1.26	0.00	0.00	1.40	1.40
65	1.20	5.16	2.58	0.00	0.00	0.00	0.52	0.52
70	0.79	3.78	4.09	0.31	0.00	0.00	1.57	1.57
75	0.24	2.44	1.95	0.24	0.00	0.00	0.24	0.24
80	1.29	3.22	2.36	0.00	0.00	0.21	0.43	0.43
85	2.32	2.53	2.11	0.00	0.00	0.00	0.21	0.21
90	2.52	4.70	1.34	0.34	0.00	0.00	0.50	0.50
95	3.77	6.13	2.59	0.47	0.00	0.00	0.47	0.47
100	2.87	4.73	2.58	0.57	0.00	0.00	0.57	0.57
105	0.87	4.88	2.44	0.00	0.00	0.00	0.87	0.87
110	5.46	3.58	3.58	0.00	0.00	0.00	0.51	0.51
115	1.38	8.06	3.54	0.20	0.00	0.00	0.20	0.20
120	0.86	4.32	4.84	0.52	0.00	0.00	1.04	1.04
125	1.39	5.35	3.76	0.59	0.00	0.00	1.78	1.78
130	1.55	2.72	2.14	0.58	0.00	0.00	1.17	1.17
135	0.92	4.78	3.12	0.18	0.00	0.00	0.37	0.37
140	1.18	4.72	1.97	0.00	0.00	0.00	0.00	0.00
145	1.43	1.02	1.64	0.41	0.00	0.00	0.82	0.82

*Liguliflorae-type and *Aster*-type
**Pollen Sum is the total counts of all terrestrial pollen (AP and NAP)

Table F.7. Percent of Aquatic Taxa* in Samples from Five Lakes

Depth (cm b.i.)	% Aquatics	% Typha	% Pediastrum	% Ruppia	% Cupressaceae
0	2.79	0.00	0.00	0.23	2.56
5	2.83	0.00	0.77	0.94	1.12
10	4.00	0.00	0.20	0.57	3.23
15	3.35	1.52	0.00	0.46	1.37
20	2.83	0.00	0.22	1.52	1.09
25	4.21	0.00	0.21	0.80	3.20
30	2.48	0.00	0.00	0.00	2.48
35	2.71	0.85	0.15	0.28	1.42
40	1.85	0.00	0.38	0.00	1.48
45	4.34	0.17	0.35	0.00	3.83
50	3.39	0.00	0.19	0.00	3.20
55	3.90	0.00	0.00	0.00	3.90
60	1.25	0.41	0.70	0.00	0.14
65	7.66	0.64	4.47	0.32	2.23
70	8.85	1.73	4.09	0.00	3.03
75	9.55	1.11	7.33	0.00	1.11
80	5.73	0.00	4.72	0.00	1.01
85	13.97	0.74	9.92	0.00	3.31
90	28.65	0.00	27.35	0.13	1.17
95	23.55	0.00	18.63	0.00	4.91
100	14.50	0.50	10.76	0.00	3.24
105	14.29	0.61	11.86	0.00	1.83
110	147.70	1.35	144.93	0.00	1.42
115	14.74	0.00	12.18	0.85	1.71
120	18.27	0.00	16.23	0.15	1.89
125	40.93	0.55	36.63	0.28	3.47
130	19.78	0.81	16.72	0.81	1.45
135	14.41	0.64	12.49	0.64	0.64
140	144.69	0.79	141.54	0.08	2.29
145	34.78	0.61	34.02	0.00	0.15

*in keeping with established palynological protocols, % aquatics are calculated by taking the count for each aquatic taxon, dividing it by the terrestrial pollen sum, and multiplying by 100

**Table F.8. Absolute Influx (# of pollen grains per year)
for Arboreal Taxa in Samples from Five Lakes**

Depth (cm b.i.)	Pinus	Larix/ Pseudo-tsuga	Acer	Betula	Alnus	Tsuga	Abies	Picea	Quercus
0	527	0	0	23	18	18	2	8	12
5	1,913	0	0	233	178	41	0	41	34
10	1,991	19	19	362	278	19	42	65	46
15	1,901	0	9	82	118	27	23	46	5
20	3,940	32	32	416	336	192	64	72	128
25	1,160	0	4	149	62	33	12	29	29
30	3,604	69	92	206	137	46	23	114	69
35	2,908	14	41	164	137	27	7	96	34
40	3,593	19	29	107	58	39	19	73	29
45	1,874	0	4	83	83	26	0	28	4
50	716	0	3	10	24	7	0	7	2
55	941	0	2	7	34	22	10	13	0
60	2,304	0	3	9	39	21	3	22	0
65	1,681	6	0	17	43	29	0	30	3
70	3,247	0	0	36	93	31	36	51	0
75	2,306	0	5	16	16	26	5	63	0
80	5,127	0	0	10	102	81	0	112	10
85	7,190	0	0	14	41	96	14	179	14
90	5,704	18	0	65	166	92	9	125	9
95	7,790	0	0	73	182	55	0	200	18
100	8,209	0	12	173	220	69	12	197	0
105	2,095	0	0	34	67	50	4	36	8
110	2,410	0	5	34	64	44.	10	62	5
115	2,606	0	13	97	97	71	0	120	0
120	1,862	0	11	38	42	34	0	63	4
125	3,736	18	18	101	110	83	64	120	18
130	2,244	0	5	25	84	44	0	42	10
135	3,802	0	8	40	87	64	0	111	16
140	3,845	0	0	8	68	93	38	98	17
145	2,716	0	6	12	29	53	6	38	0

Table F.9. Absolute Influx (# of pollen grains per year) for Non-arboreal Taxa
in Samples from Five Lakes

Depth (cm b.i.)	Cheno-podiaceae	Poaceae	Artemesia-Type	Ambrosia-Type	Poly-gonum	Apiaceae	Other Composites (Asteraceae)
0	41	27	25	20	4	4	34
5	171	260	534	41	7	0	144
10	297	603	677	93	37	19	213
15	114	1723	269	41	0	0	123
20	465	577	561	208	16	0	336
25	111	116	194	33	4	4	70
30	526	526	423	206	0	34	435
35	328	301	363	89	0	0	260
40	68	252	543	87	0	29	301
45	52	152	135	30	0	0	56
50	9	54	52	3	0	3	9
55	10	115	78	2	2	0	12
60	6	54	30	27	0	0	30
65	20	87	43	0	0	0	9
70	26	124	134	10	0	0	51
75	5	52	42	5	0	0	5
80	61	152	112	0	0	10	20
85	152	165	138	0	0	0	14
90	139	259	74	18	0	0	28
95	291	472	200	36	0	0	36
100	231	382	208	46	0	0	46
105	21	117	59	0	0	0	21
110	158	103	103	0	0	0	15
115	45	266	117	6	0	0	6
120	19	95	106	11	0	0	23
125	64	248	175	28	0	0	83
130	39	69	54	15	0	0	29
135	40	207	135	8	0	0	16
140	51	204	85	0	0	0	0
145	41	29	47	12	0	0	23

Appendix G. Charcoal Data from Five Lakes

Table G.1. Counts of Charcoal Particles for Three Size Classes

Depth (cm b.i.)	Approximate Age (^{14}C yr B.P.)*	# *Lycopodium* (aliquot)	# Charcoal Particles 75-125 µm Diameter	# Charcoal Particles 125-250 µm Diameter	# Charcoal Particles 250-500 µm Diameter
0	0	1,220	116	11	0
5	58	316	17	0	0
10	116	160	7	3	0
15	174	475	13	0	0
20	232	135	21	1	0
25	290	524	70	2	0
30	348	189	58	5	0
35	406	316	31	2	0
40	464	223	50	1	0
45	522	498	65	11	0
50	580	1,238	63	12	0
55	628	884	61	19	1
60	676	728	44	12	0
65	724	749	52	16	0
70	772	420	46	17	3
75	820	413	36	7	0
80	866	213	83	11	2
85	912	157	79	5	0
90	958	234	83	15	1
95	1,004	119	91	17	1
100	1,050	187	60	6	1
105	1,105	516	85	16	1
110	1,160	439	73	23	2
115	1,215	333	79	24	1
120	1,270	569	121	39	2
125	1,325	235	134	32	2
130	1,378	440	30	14	0
135	1,431	272	27	3	0
140	1,484	255	41	22	0
145	1,537	369	48	21	4

*determined via linear interpolation from AMS dates reported in Tables 4.2 and 4.3

Table G.2. Absolute Influx of Charcoal Particles in #/cm²/yr

Depth (cm b.i.)	Approximate Age (^{14}C yr B.P.)*	# *Lycopodium* (aliquot)	# Charcoal Particles 75-125 μm Diameter	# Charcoal Particles 125-250 μm Diameter	# Charcoal Particles 250-500 μm Diameter
0	0	225.10	205.61	19.50	0.00
5	58	30.13	30.13	0.00	0.00
10	116	17.72	12.41	5.32	0.00
15	174	23.04	23.04	0.00	0.00
20	232	38.99	37.22	1.77	0.00
25	290	127.62	124.07	3.54	0.00
30	348	111.67	102.80	8.86	0.00
35	406	58.49	54.95	3.54	0.00
40	464	90.40	88.62	1.77	0.00
45	522	134.71	115.21	19.50	0.00
50	580	132.94	111.67	21.27	0.00
55	628	143.57	108.12	33.68	1.77
60	676	99.26	77.99	21.27	0.00
65	724	120.53	92.17	28.36	0.00
70	772	116.98	81.53	30.13	5.32
75	820	76.22	63.81	12.41	0.00
80	866	170.16	147.12	19.50	3.54
85	912	148.89	140.03	8.86	0.00
90	958	175.47	147.12	26.59	1.77
95	1,004	193.20	161.29	30.13	1.77
100	1,050	118.76	106.35	10.63	1.77
105	1,105	180.79	150.66	28.36	1.77
110	1,160	173.70	129.39	40.77	3.54
115	1,215	184.34	140.03	42.54	1.77
120	1,270	287.14	214.47	69.13	3.54
125	1,325	297.78	237.51	56.72	3.54
130	1,378	77.99	53.17	24.81	0.00
135	1,431	53.17	47.86	5.32	0.00
140	1,484	111.67	72.67	38.99	0.00
145	1,537	129.39	85.08	37.22	7.09

*determined via linear interpolation from AMS dates reported in Tables 4.2 and 4.3

Table G.3. Absolute Influx of Charcoal Particles in mm^2/cm^2/year

Depth (cm b.i.)	Approximate Age (^{14}C yr B.P.)*	# *Lycopodium* (aliquot)	# Charcoal Particles 75-125 μm Diameter	# Charcoal Particles 125-250 μm Diameter	# Charcoal Particles 250-500 μm Diameter
0	0	2.74	2.06	0.69	0.00
5	58	0.30	0.30	0.00	0.00
10	116	0.31	0.12	0.19	0.00
15	174	0.23	0.23	0.00	0.00
20	232	0.43	0.37	0.06	0.00
25	290	1.37	1.24	0.12	0.00
30	348	1.34	1.03	0.31	0.00
35	406	0.67	0.55	0.12	0.00
40	464	0.95	0.89	0.06	0.00
45	522	1.84	1.15	0.69	0.00
50	580	1.86	1.12	0.75	0.00
55	628	2.51	1.08	1.18	0.25
60	676	1.53	0.78	0.75	0.00
65	724	1.92	0.92	1.00	0.00
70	772	2.62	0.82	1.06	0.75
75	820	1.07	0.64	0.44	0.00
80	866	2.66	1.47	0.69	0.50
85	912	1.71	1.40	0.31	0.00
90	958	2.66	1.47	0.93	0.25
95	1,004	2.92	1.61	1.06	0.25
100	1,050	1.69	1.06	0.37	0.25
105	1,105	2.75	1.51	1.00	0.25
110	1,160	3.23	1.29	1.43	0.50
115	1,215	3.15	1.40	1.50	0.25
120	1,270	5.07	2.14	2.43	0.50
125	1,325	4.87	2.38	1.99	0.50
130	1,378	1.40	0.53	0.87	0.00
135	1,431	0.67	0.48	0.19	0.00
140	1,484	2.10	0.73	1.37	0.00
145	1,537	3.16	0.85	1.31	1.00

*determined via linear interpolation from AMS dates reported in Tables 4.2 and 4.3

Appendix H. Proxies Taken at 50 cal-year Intervals from cal A.D. 950 to 1900

Table H.1. Variables used in Correlations and Time Series Analysis: Composite Isotope Index, Human Population Index, and Charcoal, Reported at 50 year intervals

Date A.D.	Temporal Interval A.D.	Oxygen Isotope Index	Human Population Index	Influx of Large Charcoal*
1900	1875-1925	-1.23	-1.58	3.85
1850	1825-1875	-0.89	-1.35	2.20
1800	1775-1825	-0.55	0.00	0.79
1750	1725-1775	-0.25	-0.23	2.32
1700	1675-1725	0.04	-0.54	4.46
1650	1625-1675	0.24	-0.60	8.68
1600	1575-1625	0.44	-0.60	4.10
1550	1525-1575	0.22	-0.60	2.20
1500	1475-1525	0.00	-0.36	12.77
1450	1425-1475	-0.58	0.36	20.35
1400	1375-1425	-1.16	0.41	27.18
1350	1325-1375	-0.90	0.64	28.95
1300	1275-1325	-0.63	1.17	24.81
1250	1225-1275	-0.14	0.44	32.20
1200	1175-1225	0.36	0.72	22.01
1150	1125-1175	0.59	0.20	19.34
1100	1075-1125	0.83	0.22	12.56
1050	1025-1075	1.02	0.35	24.97
1000	975-1025	1.21	0.71	31.60
950	925-975	1.39	0.65	12.41

*number of particles $\geq$ 125 microns in diameter deposited per square centimeter per year

**Table H.2. Variables used in Correlations and Time Series Analysis:
Non-arboreal Pollen reported at 50-year intervals**

Temporal Interval A.D.	Interval Midpoint cal A.D.	% NAP	% Cheno-podiaceae	% Poaceae	% Artemesia-type	% Ambrosia-type	% Aster-type	% Polygon-aceae	% Liguli-florae-type	% Apiaceae
1875-1925	1900	37.19	5.92	11.34	14.61	1.75	2.52	0.63	0.14	0.29
1825-1875	1850	30.07	4.93	8.84	11.43	1.65	2.37	0.33	0.36	0.16
1775-1825	1800	25.18	5.08	6.91	8.64	2.08	2.11	0.10	0.26	0.00
1725-1775	1750	26.64	6.22	7.36	8.44	2.52	1.75	0.22	0.06	0.06
1675-1725	1700	26.23	6.10	6.28	9.28	1.95	1.98	0.17	0.20	0.27
1625-1675	1650	30.71	8.30	8.28	6.75	3.22	3.46	0.00	0.18	0.53
1575-1625	1600	27.18	7.16	6.63	7.64	2.04	3.63	0.00	0.02	0.06
1525-1575	1550	24.00	2.69	5.26	9.86	1.74	4.02	0.00	0.00	0.43
1475-1525	1500	18.59	1.79	5.63	7.34	1.40	2.11	0.00	0.11	0.21
1425-1475	1450	15.04	1.55	6.06	5.61	0.82	0.73	0.00	0.09	0.19
1375-1425	1400	16.64	0.94	7.93	6.41	0.32	0.72	0.09	0.00	0.23
1325-1375	1350	13.42	0.62	6.89	4.55	0.70	0.55	0.12	0.00	0.00
1275-1325	1300	7.53	0.74	3.84	1.99	0.63	0.33	0.00	0.00	0.00
1225-1275	1250	9.88	0.98	4.41	3.40	0.17	0.83	0.00	0.09	0.00
1175-1225	1200	7.11	0.47	3.00	2.85	0.27	0.46	0.00	0.07	0.00
1125-1175	1150	6.59	0.92	2.95	2.22	0.08	0.28	0.00	0.00	0.14
1075-1125	1100	7.26	2.05	2.71	2.17	0.00	0.11	0.00	0.16	0.06
1025-1075	1050	8.73	2.48	4.32	1.48	0.28	0.14	0.00	0.04	0.00
975-1025	1000	12.63	3.66	6.01	2.49	0.46	0.01	0.00	0.00	0.00
925-975	950	10.76	2.87	4.73	2.58	0.57	0.00	0.00	0.00	0.00

**Table H.3. Variables used in Correlations and Time Series Analysis:
Arboreal Pollen reported at 50-year intervals**

Temporal Interval A.D.	Interval Midpoint cal A.D.	% AP	% Pinus	% Larix/ Pseudo-tsuga	% Acer	% Betula	% Alnus	% Tsuga	% Abies	% Picea	% Quercus
1875-1925	1900	62.81	45.60	0.29	0.29	7.40	5.68	0.61	0.65	1.32	0.98
1825-1875	1850	69.93	56.15	0.16	0.35	4.86	4.86	0.72	0.83	1.50	0.50
1775-1825	1800	74.82	60.95	0.20	0.37	4.17	4.36	1.72	0.84	1.32	0.89
1725-1775	1750	73.36	56.12	0.31	0.37	6.34	4.21	2.37	0.81	1.15	1.69
1675-1725	1700	73.77	58.41	0.19	0.42	6.78	2.97	1.51	0.58	1.52	1.40
1625-1675	1650	69.29	57.34	1.06	1.43	3.28	2.20	0.72	0.36	1.82	1.08
1575-1625	1600	72.82	61.62	0.37	0.94	3.48	2.85	0.60	0.17	2.02	0.77
1525-1575	1550	76.00	67.83	0.36	0.64	2.42	1.56	0.71	0.32	1.56	0.60
1475-1525	1500	81.41	73.00	0.14	0.32	2.84	2.48	0.93	0.14	1.24	0.32
1425-1475	1450	84.96	77.32	0.00	0.28	2.27	3.02	0.92	0.00	0.96	0.18
1375-1425	1400	83.36	76.42	0.00	0.32	0.96	2.90	1.30	0.38	0.97	0.11
1325-1375	1350	86.58	80.04	0.00	0.19	0.56	2.54	1.55	0.55	1.15	0.00
1275-1325	1300	92.47	86.37	0.17	0.07	0.73	2.20	1.35	0.07	1.43	0.09
1225-1275	1250	90.12	82.52	0.16	0.00	1.07	2.72	1.30	0.60	1.68	0.08
1175-1225	1200	92.89	86.18	0.00	0.14	0.89	1.61	1.11	0.60	2.37	0.00
1125-1175	1150	93.41	86.95	0.00	0.08	0.39	1.65	1.54	0.08	2.56	0.14
1075-1125	1100	92.74	86.95	0.00	0.00	0.21	1.03	1.54	0.16	2.64	0.21
1025-1075	1050	91.27	83.04	0.28	0.00	1.01	2.61	1.64	0.18	2.35	0.18
975-1025	1000	87.37	80.36	0.03	0.00	0.96	2.42	0.79	0.01	2.57	0.23
925-975	950	89.24	80.77	0.00	0.14	2.15	2.73	0.86	0.14	2.44	0.00

References Cited

Abrams, M.D.
2005   Prescribing Fire in Eastern Oak Forests: Is Time Running Out? *Northern Journal of Applied Forestry* 22(3):190-196.

Abrams, M.D., and G.J. Nowacki
2008   Native Americans as Active and Passive Promoters of Mast and Fruit Trees in the Eastern USA. *The Holocene* 18(7):1123-1137.

Abrams, E.M., and D.J. Rue
1988   The Causes and Consequences of Deforestation among the Prehistoric Maya. *Human Ecology* 16(4):377-395.

Aikens, M.C.
1988   Review [Untitled]. *American Antiquity* 53(3):659-661.

Albert, B.M.
2007   Climate, Fire, and Land-use History in the Oak-Pine-Hickory Forests of Northeast Texas during the Past 3500 Years. *Castanea* 72:82-91.

Allen, C.D.
2002   Lots of Lightning and Plenty of People: An Ecological History of Fire in the Upland Southwest. In *Native Peoples and the Natural Landscape*, edited by T.R. Vale, pp. 143-193. Island Press, Washington, D.C.

Alt, D., and D.W. Hyndman
1995   *Northwest Exposures: A Geologic Story of the Northwest*. Mountain Press Publishing, Missoula, MT.

Ames, K.M.
2000   *Review of the Archaeological Data* (Chapter 2 in the Cultural Affiliation Report). Data retrieved from the National Park Service web site on Kennewick Man: http://www.cr.nps.gov/aad/kennewick/ames.htm

Ames, K.M., D.D. Dumond, J.R. Galm, and R. Minor
1998   Prehistory of the Southern Plateau. In *Plateau*, edited by D.E. Walker Jr., pp. 103-119. Handbook of North American Indians, Volume 12, W.C. Sturtevant, general editor, Smithsonian Institution, Washington D.C.

Arno, S.F. and K.M. Sneck
1977   A Method for Determining Fire History in Coniferous Forests of the Mountain West. *USDA Forest Service General Technical Report INT-42*. Intermountain Forest and Range Experiment Station. Forest Service, U.S. Department of Agriculture, Ogden, Utah.

Aschmann, H.
1977   Aboriginal Use of Fire. In *Proceedings of the Symposium on the Environmental Consequences of Fire and Fuel Management in Mediterranean Ecosystems*, (H.A. Mooney and C.E. Conrad, technical coordinators), pp. 132-141. General Technical Report WO-3, U.S. Department of Agriculture, Forest Service, Washington D.C.

Athens, J.S. and J.V. Ward
1993   Environmental Change and Prehistoric Polynesian Settlement in Hawai'i. *Asian Perspectives* 32:205-223.

Backman, A.E.
1984   *1000-year Record of Fire-Vegetation Interactions in the Northeastern United States: A Comparison of Coastal and Inland Regions*. Master's Thesis, University of Massachusetts, Amherst.

Baker, W.L.
2002   Indians and Fire in the Rocky Mountains: The Wilderness Hypothesis Renewed. In *Native Peoples and the Natural Landscape*, edited by T.R. Vale, pp. 41-76. Island Press, Washington, D.C.

Baker, W.L., and D. Ehle
2001   Uncertainty in Surface-fire History: The Case of Ponderosa Forests in the Western United States. *Canadian Journal of Forest Research* 31:1205-1226.

Barrett, S., and S. Arno
1999   Indian Fires in the Northern Rockies. In *Indians, Fire, and the Land in the Pacific Northwest*, edited by R. Boyd, pp. 50-64. Oregon State University Press, Corvallis.

Bean, L.J., and H.W. Lawton
1993   Some Explanations for the Rise of Cultural Complexity in Native California with Comments on Proto-Agriculture and Agriculture. In *Before the Wilderness: Environmental Management by Native Californians*, edited by T.C. Blackburn and K. Anderson, pp. 27-54. Ballena Press, Menlo Park, California. (Originally published in 1973 as Ballena Press Anthropological Papers 1).

Behling, H. and M.L. da Costa
2000   Holocene Environmental Changes from the Rio Curuá Record in the Caxiuanã Region, Eastern Amazon Basin. *Quaternary Research* 53(3):369-377.

Bendix, J.
2002 Pre-European Fire in California Chaparral. In *Native Peoples and the Natural Landscape*, edited by T.R. Vale, pp. 143-193. Island Press, Washington, D.C.

Bense, J.A.
1972 *The Cascade Phase: A Study in the Effect of the Altithermal on a Cultural System*. Ph.D. Dissertation, Washington State University. Pullman, WA.

Bergeron, Y., and S. Archambault
1993 Decreasing Frequency of Forest Fires in the Southern Boreal Zone of Québec and its Relation to Global Warming Since the End of the "Little Ice Age." *The Holocene* 3(3):255-259.

Bernabo, J.C.
1981 Quantitative Estimates of Temperature Changes over the Last 2700 Years in Michigan Based on Pollen Data. *Quaternary Research* 15:143-159.

Binford, M.W., T.J. Whitmore, A. Higuera-Gundy, E.S. Deevey Jr., and B. Leyden
1987 Ecosystems, Paleoecology, and Human Disturbance in Subtropical and Tropical America. *Quaternary Science Reviews* 6(2):115-128.

Birks, H.J.B.
1981 The Use of Pollen Analysis in the Reconstruction of Past Climates: A Review. In *Climate and History*, edited by T.M.L. Wigley, M.J. Ingram and G. Farmer, pp. 111-138. Cambridge University Press, Cambridge.

Birks, H.J.B., and H.H. Birks
1980 *Quaternary Palaeoecology*. Edward Arnold, London

Birks, H.J.B., and A.D. Gordon
1985 *Numerical Methods in Quaternary Pollen Analysis*. Academic Press, London.

Black, J.
1970 *The Dominion of Man: The Search for Ecological Responsibility*. Edinburgh University Press, Edinburgh.

Black, B.A., C.M. Ruffner, and M.D. Abrams
2006 Native American Influences on the Forest Composition of the Allegheny Plateau, Northwest Pennsylvania. *Canadian Journal of Forest Research* 36:1266-1275.

Blackburn, T., and K. Anderson
1993 Introduction: Managing the Domesticated Environment. In *Before the Wilderness: Environmental Management by Native Californians*, edited by T.C. Blackburn and K. Anderson, pp. 15-24. Ballena Press, Menlo Park, California.

Boyd, M.
2002 Identification of Anthropogenic Burning in the Paleoecological Record of the Northern Prairies: A New Approach. *Annals of the Association of American Geographers* 92(3):471–487.

Boyd, R.
1985 *The Introduction of Infectious Diseases Among the Indians of the Pacific Northwest, 1774-1874*. Ph.D. Dissertation, University of Washington, Seattle.
1990 Demographic History, 1774-1874. In *Northwest Coast*, edited by W.P Suttles, pp. 135-148. Handbook of North American Indians, Volume 7, W.C.S. Sturtevant, general editor, Smithsonian Institution, Washington D.C.
1999 Introduction. In *Indians, Fire, and The Land in the Pacific Northwest*, edited by R. Boyd, pp. 1-30. Oregon State University Press, Corvallis.

Bradley, R.S.
1999 *Paleoclimatology : Reconstructing Climates of the Quaternary*. Academic Press, San Diego.

Bradley, R.S., and P.D. Jones
1993 Little Ice Age Summer Temperature Variations: their Nature and Relevance to Recent Global Warming Trends. *The Holocene* 3(4):367-376.

Bradshaw, R.H.W.
1981 Quantitative Reconstruction of Local Woodland Vegetation using Pollen Analysis from a Small Basin in Norfolk, England. *Journal of Ecology* 69(3):941-955.

Broeker, W. S.
2001 Was the Medieval Warm Period Global? *Science* 291(5508):1497-1499.

Broughton, J.M.
1994 Late Holocene Resource Intensification in the Sacramento Valley, California; The Vertebrate Evidence. *Journal of Archaeological Science* 21(4):501-514.
1995 *Resource Depression and Intensification During the Late Holocene, San Francisco Bay: Evidence from the Emeryville Shellmound Vertebrate Fauna*. Ph.D. Dissertation, University of Washington, Seattle.

Brown, A.G.
1997 *Alluvial Geoarchaeology; Floodplain Archaeology and Environmental Change*. Cambridge Manuals in Archaeology, Cambridge University Press, Cambridge.
1999 Characterising Prehistoric Lowland Environments using Local Pollen Assemblages. *Journal of Quaternary Science* 14(6):585-594.

Brown, K.J., and R.J. Hebda
2002a Ancient Fires on Southern Vancouver Island, British Columbia, Canada: A Change in Causal Mechanisms at about 2000 ybp. *Environmental Archaeology* 7:1-12.
2002b Origin, Development, and Dynamics of Coastal Temperate Conifer Rainforests of Southern Vancouver Island, Canada. *Canadian Journal of Forest Research* 32:353-372.

Brown, T.A., D.E. Nelson, R.W. Mathewes, J.S. Vogel, and J.R. Southon
1989 Radiocarbon Dating of Pollen by Accelerator Mass Spectrometry. *Quaternary Research* 32:205-212.

Buckland, P.D., and K.J. Edwards
1998 Palaeoecological Evidence for Possible Pre-European Settlement in the Falkland Islands. *Journal of Archaeological Science* 25:599-602.

Buckner, E.
2000 Summary: Fire in the Evolution of the Eastern Landscape—A Timeline. In *Proceedings: Workshop on Fire, People and the Central Hardwoods Landscape. March 12-14 2000 Richmond, Kentucky.* USDA Forest Service Northeastern Research Station General Technical Report NE-274, p. 12.

Burcham, L.T.
1974 Fire and Chaparral before European Settlement. In *Symposium on Living with the Chaparral; Proceedings*, edited by M. Rosenthal, pp. 101-120. Sierra Club, San Francisco.

Burney, D.A.
1993 Late Holocene Environmental Changes in Arid Southwestern Madagascar. *Quaternary Research* 40:98-106.

Burney, D.A., G.S. Robinson, and L.P. Burney
2003 *Sporormiella* and the Late Holocene Extinctions in Madagascar. *Proceedings of the National Academy of Sciences* 100(19):10800-10805.

Cairns, J., Jr.
1980 *The Recovery Process in Damaged Ecosystems.* Ann Arbor Science Publishers Inc., Ann Arbor.

Campbell, S.K.
1985 *Summary Report, Chief Joseph Dam Cultural Resources Project.* Office of Public Archaeology, University of Washington, Seattle.
1989 *Post-Columbian Culture History in the Northern Columbia Plateau: A.D. 1500-1900.* Ph.D. Dissertation, University of Washington, Seattle, WA.
1990 *Post Columbian Culture History in the Northern Columbia Plateau A.D. 1500-1900.* Garland Publishing Inc., New York.

Campbell, I.D. and J.H. McAndrews

1995 Charcoal Evidence for Indian-set Fires: A Comment on Clark and Royall. *The Holocene* 5(3):369-370.

Carcaillet, C., M. Bouvier, B. Fréchette, A.C. Larouche, and P.J.H. Richard
2001 Comparison of Pollen-slide and Sieving Methods in Lacustrine Charcoal Analyses for Local and Regional Fire History. *The Holocene* 11(4):467-476.

Casselberry, S.E.
1974 Further Refinement of Formulae for Determining Population from Floor Area. *World Archaeology* 6(1):117-122.

Chapman, J., H.R. Delcourt, and P.A. Delcourt
1989 Strawberry Fields, Almost Forever. *Natural History* 9:50-58.

Chatters, J.C.
1995a *Resource Intensification and Demography on the Columbian Plateau of Western North America.* Paper delivered to the Society for American Archaeology conference, Minneapolis, MN. May 6, 1995.
1995b Population Growth, Climatic Cooling, and the Development of Collector Strategies on the Southern Plateau, Western North America. *Journal of World Prehistory* 9(3):341-400.
1998 Environment. In *Plateau*, edited by D.E. Walker Jr., pp. 29-48. Handbook of North American Indians, Volume 12, W.C. Sturtevant, general editor, Smithsonian Institution, Washington D.C.

Chatters, J.C, and D. Leavell
1994 *Smeads Bench Bog: A 1400 Year History of Fire and Succession in the Hemlock Forest of the Lower Clark Fork Valley, Northwestern Montana.* Unpublished manuscript in possession of the authors, prepared as a report for Kootenai National Forest.

Clark, J.S.
1988a Effect of Climate Change on Fire Regimes in Northwestern Minnesota. *Nature* 334:233-235.
1988b Particle Motion and the Theory of Charcoal Analysis: Source Area, Transport, Deposition, and Sampling. *Quaternary Research* 30:81-91.
1988c Charcoal—Stratigraphic Charcoal Analysis on Petrographic Thin Sections: Recent Application of Fire History in Northwestern Minnesota. *Quaternary Research* 30:67-80.
1995 Climate and Indian Effects on Southern Ontario Forests: A Reply to Campbell and McAndrews. *The Holocene* 5(3):371-379.

Clark, J.S. and T.C. Hussey
1996 Estimating the Mass Flux of Charcoal from Sedimentary Records: Effects of Particle Size, Morphology and Orientation. *The Holocene* 6(2):129-144.

Clark, J.S. and P.D. Royall
  1995a Transformation of a Northern Hardwood Forest by Aboriginal (Iroquois) Fire: Charcoal Evidence from Crawford Lake, Ontario, Canada. *The Holocene* 5(1):1-9.
  1995b Particle-size Evidence for Source Areas of Charcoal Accumulation in Late Holocene Sediments of Eastern North American Lakes. *Quaternary Research* 43:80-89.
  1996 Local and Regional Sediment Charcoal Evidence for Fire Regimes in Presettlement North-eastern North America. *The Journal of Ecology* 84(3):365-382.

Clark, R.L.
  1984 Effects on Charcoal of Pollen Preparation Procedures. *Pollen et Spores* 26:559-576.
  1987 Fire History from Charcoal: Effects of Sampling on Interpretation. In *Archaeometry: further Australasian Studies*, edited by W.R. Ambrose and J.M.J. Mummery, pp. 135-142. Australian National Gallery, Canberra.

Cole, K.L. and G.W. Liu
  1994 Holocene Paleoecology of an Estuary on Santa Rosa Island, California. *Quaternary Research* 41(3):326-335.

Cook, S.F.
  1976 *The Population of the California Indians, 1769-1970*. University of California Press, Berkeley.

Cook, S.F., and A.E. Treganza
  1950 The Quantitative Investigation of Indian Mounds with Special Reference to the Relation of the Physical Components to the Probable Material Culture. *University of California Publication of American Archaeology and Ethnology* 40:223-262.

Cox, T.R.
  1999 Changing Forests, Changing Needs: Using the Pacific Northwest's Westside Forests, Past and Present. In *Northwest Lands, Northwest Peoples*, edited by D.D. Goble and P.W. Hirt, pp. 462-476. University of Washington Press, Seattle.

Craig, H., L.I. Gordon, and Y. Horibe
  1963 Isotopic Exchange Effects in the Evaporation of Water. *Journal of Geophysical Research* 68(17):5079-5087.

Cromwell, J.B., M.J. Hannan, W.C. Labys, W.C., and M. Terraza
  1994 *Multivariate Tests for Time Series Models*. Sage Publications, Thousand Oaks, CA.

Crowley, T.J., and T.S. Lowery
  2000 How Warm was the Medieval Warm Period? *Ambio* 29(1):51-55.

Cushing, E.J.
  1967 Evidence for Differential Pollen Preservation in Late Quaternary Sediments in Minnesota. *Review of Palaeobotany and Palynology* 4:87-101.

Cwynar, L. C.
  1978 Recent History of Fire and Vegetation from Laminated Sediment of Greenleaf Lake, Algonquin Park, Ontario. *Canadian Journal of Botany* 56:10-21.
  1987 Fire and the Forest History of the North Cascade Range. *Ecology* 68(4):791-802.

Dalan, R.
  1985a Pollen Analysis of a Core from Goose Lake, Okanogan County, Washington. In *Summary of Results, Chief Joseph Dam Cultural Resources Project*, edited by S.K. Campbell, pp. 113-129. Office of Public Archaeology, University of Washington, Seattle.
  1985b Appendix C: Pollen Analysis of a core from Rex Grange Lake, Douglas County, Washington. In *Summary Report, Chief Joseph Dam Cultural Resources Project*, edited by S.K. Campbell, pp. 521-527. Office of Public Archaeology, University of Washington, Seattle.
  1993 Issues of Scale in Archaeogeophysical Research. In *Effects of Scale on Archaeological and Geoscientific Perspectives*, edited by J.K. Stein, J.K. and A.R. Linse, pp. 67-78. Geological Society of America Special Paper 283. Boulder, Co.

Dansgaard, W.
  1964 Stable Isotopes in Precipitations. *Tellus* 16:436-468.

Daubenmire, R.
  1970 *Steppe Vegetation of Washington*. Washington Agricultural Experiment Station Technical Bulletin 62.

Daubenmire, R., and J.B. Daubenmire
  1968 *Forest Vegetation of Eastern Washington and Idaho*. Washington Agricultural Experiment Station Technical Bulletin 60.

Davis, M.B., R.E. Moeller, and J. Ford
  1984 Sediment Focusing and Pollen Influx. In *Lake Sediments and Environmental History*, edited by E.Y. Haworth and J.W.G. Lund, pp. 261-293, University of Leicester Press, Leicester.

Davis, O.K.
  1987 Spores of the Dung Fungus *Sporormiella*: Increased Abundance in Historic Sediments and before Pleistocene Megafaunal Extinction. *Quaternary Research*. 28(2): 290-294.

Davis, O.K., D.A. Kova and P.J. Mehringer Jr.
  1977 Pollen Analysis of Wildcat Lake, Whitman County, Washington: The Last 1000 Years. *Northwest Science* 51(1):13-30.

Davis , O.K., and D.S. Shafer
2006   *Sporormiella* Fungal Spores, a Palynological Means of Detecting Herbivore Density. *Palaeogeography, Palaeoclimatology, Palaeoecology* 237(1):40-50

Day, G.M.
1953   The Indian as an Ecological Factor in the Northeastern Forest. *Ecology* 34:329-346.

D'Costa, D.M., J. Grindrod, R. Ogden
1993   Preliminary Environmental Reconstructions from Late Quaternary Pollen and Mollusc Assemblages at Egg Lagoon, King Island, Bass Strait. *Austral Ecology* 18(3):351-366.

Dean, W.E., Jr.
1974   Determination of Carbonate and Organic Matter in Calcareous Sediments and Sedimentary Rocks by Loss on Ignition: Comparison with Other Methods. *Journal of Sedimentary Petrology* 44:242-248.

Delcourt, H.R.
1987   The Impact of Prehistoric Agriculture and Land Occupation on Natural Vegetation. *Trends in Ecology and Evolution* 2(2):39-44.

Delcourt H.R. and P.A. Delcourt
1985   Quaternary Palynology and Vegetational History of the Southeastern United States. In *Pollen Records of Late-Quaternary North American Sediments*, edited by V.M. Bryant Jr., and R.G. Holloway, pp. 1-3. American Association of Stratigraphic Palynologists Foundation, Dallas.
1988   Quaternary Landscape Ecology: Relevant Scales in Space and Time. *Landscape Ecology* 2:23-44.
1991   *Quaternary Ecology: A Paleoecological Perspective*. Kluwer Academic Publication, New York.

Delcourt, P.A. and H.R. Delcourt
1987a *Long-term Forest Dynamics of the Temperate Zone: A Case Study of Late-Quaternary Forests in Eastern North America*. Springer-Verlag, New York.
1987b Late-Quaternary Dynamics of Temperate Forests: Applications of Paleoecology to Issues of Global Environmental Change. *Quaternary Science Reviews* 6(2):129-146.
1998   The Influence of Prehistoric Human-set Fires on Oak-Chestnut Forests in the Southern Appalachians. *Castanea* 63(3): 337-345.
2004   *Prehistoric Native Americans and Ecological Change: Human Ecosystems in Eastern North America since the Pleistocene*. Cambridge University Press. Cambridge.

Delcourt, P.A., H.R. Delcourt, C.R. Ison, W.E. Sharp, and K.J. Gremillion
1998   Prehistoric Human Use of Fire, the Eastern Agricultural Complex, and Appalachian Oak-Chestnut Forests: Paleoecology of Cliff Palace Pond, Kentucky. *American Antiquity* 63(2):263-278.

Delcourt, H.R., P.A. Delcourt and T. Webb III
1983   Dynamic Plant Ecology: The Spectrum of Vegetational Change in Space and Time. *Quaternary Science Reviews* 1:153-175.

Dell, J.D.
1977   *Action Plan for Improving Slash Burning Programs on Westside National Forests in Region 6*. Forest Service USDA, Pacific Northwest Region.

Denevan, W.M.
1992   The Pristine Myth: The Landscape of the Americas in 1492. *Annals of the Association of American Geographers* 82(3):369-385.
1996   Carl Sauer and Native American Population Size. *Geographical Review* 86(3):385-397

Deur, D.
1999   Salmon, Sedentism, and Cultivation: Toward an Environmental Prehistory of the Northwest Coast. In *Northwest Lands and Northwest Peoples*, edited by D.D. Goble and P.W. Hirt, pp. 129-155. University of Washington Press, Seattle.

Diamond, J.
1995   Easter's End. *Discover* 16(8):62-69.
2005   *Collapse: How Societies Choose to Fail or Succeed*. Viking, New York.

Dobyns, H.
1963   Indian Extinction in the Middle Santa Cruz Valley, Arizona. *New Mexico Historical Review* 38:163-181.
1983   *Their Number Became Thinned*. The University of Tennessee Press, Knoxville.

Dorney, C.H., and J.R. Dorney
1989   An Unusual Oak Savanna in Northeastern Wisconsin: The Effect of Indian-caused Fire. *American Midland Naturalist* 122:103-113.

Dunnell, R.C.
1991   Methodological Impacts of Catastrophic Depopulation on American Archaeology and Ethnology. In *Columbian Consequences*, vol. 1, edited by D.H. Thomas, pp. 561-580. Smithsonian Institution Press, Washington.

Ellis, R.
1980   *Fuels and Fire Frequency: An Evaluation of the Role of Wildland Fuel Characteristics in Determining Fire Frequency*. Master's Thesis, University of Washington, Seattle, WA.

Emiliani, C.
1955   Pleistocene Temperatures. *Journal of Geology* 63:538-578.

Epstein, S.R., H. Buchsbaum, H.A. Lowenstam, and H.C. Urey
1953   Revised Carbonate-Water Isotopic Temperature Scale. *Bulletin of the Geological Society of America* 64:1315-1326.

Faegri, K. and J. Iverson
1989   *Textbook of Pollen Analysis.* Wiley, Chichester.

Fall, P.L.
1997   Fire History and Composition of the Subalpine Forest of Western Colorado During the Holocene. *Journal of Biogeography* 24:309-325.

Fiacco, R.J., Jr., J.M. Palais, M.S. Germani, G.A. Zielinski, and P.A. Mayewski
1993   Characteristics and Possible Source of a 1479 A.D. Volcanic Ash Layer in a Greenland Ice Core. *Quaternary Research* 39:267-273.

Flenley, J.R., and S.M. King
1984   Late Quaternary Pollen Records from Easter Island. *Nature* 307:47-50.

Flenley, J.R., A.S.M. King, J. Jackson, C. Chew, J. Teller, and M.E. Prentice
1991   The Late Quaternary Vegetational and Climatic History of Easter Island. *Journal of Quaternary Science* 6(2):85-115.

Forbes, J.
1987   *Carbon and Oxygen Isotopic Composition of Holocene Lake Sediments from Okanogan County, Washington.* Master's Thesis, University of Washington. Seattle, WA.

Foster, D.R.
2000   From Bobolinks to Bears: Interjecting Geographical History into Ecological Studies, Environmental Interpretation, and Conservation Planning. *Journal of Biogeography* 27(1):27-30.

Franklin, J.F., and C.T. Dyrness
1973   Natural Vegetation of Oregon and Washington. *USDA Forest Service Technical Report PNW-80.* Portland, Oregon.

Fritts, H.C., and X.M. Shao
1992   Mapping Climate Using Tree-rings from Western North America. In *Climate Since A.D. 1500*, edited by R.S. Bradley and P.D. Jones, pp. 269-295. Routledge, New York.

Fritz, S.C., D.R. Engstrom and B.J. Haskell
1994   'Little Ice Age' Aridity in the North American Great Plains: A High-resolution Reconstruction of Salinity Fluctuations from Devil's Lake, North Dakota, USA. *The Holocene* 4(1):69-73.

Fritz, P., and S. Poplawski
1974 ^{18}O and ^{13}C in the Shells of Freshwater Molluscs and their Environments. *Earth, and Planetary Science Letters* 24:91-98

Fulé, P.Z., T.A. Heinlein, W.W. Covington, and M.M. Moore.
2003   Assessing Fire Regimes on Grand Canyon Landscapes with Fire-scar and Fire-record Data. *International Journal of Wildland Fire* 12:129-145.

Gajewski, K., M.G. Winkler, and A.M. Swain
1985   Vegetation and Fire History from Three Lakes with Varved Sediments in Northwestern Wisconsin. *Review of Palaeobotany and Palynology* 44:277-292.

Galm, J.
1994   Prehistoric Trade and Exchange in the Interior Plateau of Northwestern North America. In *Prehistoric Exchange Systems in North America*, edited by T.G. Baugh and J.E. Ericson, pp. 275-305. Plenum Press, New York.

Gardner, J.J., and C. Whitlock, C.
2001   Charcoal Accumulation Following a Recent Fire in the Cascade Range, Northwestern USA, and its Relevance for Fire-history Studies. *The Holocene* 11(5):541-549.

Gauch, H.G.
1982   *Multivariate Analysis in Community Ecology.* Cambridge University Press, Cambridge.

Gavin, D.G., D.J. Hallett, F.S. Hu, K.P. Lertzman, S.J. Prichard, K.J. Brown, J.A. Lynch, P.J. Bartlein, and D.L. Peterson.
2007   Forest Fire and Climate Change in Western North America: Insights from Sediment Charcoal Records. *Frontiers in Ecology and the Environment* 5:499-506.

Gerhardt, F., and D.R. Foster
2002   Physiographical and Historical Effects on Forest Vegetation in Central New England, USA. *Journal of Biogeography*, 29(10-11):1421-1437.

Gottman, J.M.
1981   *Time-series Analysis: A Comprehensive Introduction for Social Scientists.* Cambridge University Press, Cambridge.

Goudie, A.
2005   *The Human Impact on the Natural Environment. Past, Present, Future.* Wiley-Blackwell, Malden, MA.

Goudsblom, J.
1992   *Fire and Civilization.* Penguin Press, London.

Granger, C.W.J.
1969   Investigating Causal Relations by Econometric Methods of Cross-Spectral Methods. *Econometrica* 34:424-438.

Graumlich, L.J., and L.B. Brubaker
1986   Reconstruction of Annual Temperatures (1590-

1979) for Longmire, Washington, Derived from Tree Rings. *Quaternary Research* 5:223-234.

Gray, J.
1965 Extraction Techniques. In *Handbook of Paleontological Techniques*, edited by B.G. Kummel and D.M. Raup, pp. 530-587. Freeman, San Francisco.

Grayson, D.K.
1991 Late Pleistocene Mammalian Extinctions in North America: Taxonomy, Chronology, and Explanations. *Journal of World Prehistory* 5(3):193-232.
2001 The Archaeological Record of Human Impacts on Animal Populations. *Journal of World Prehisto*ry 15(1):1-68

Grayson, D.K., and M.D. Cannon
1999 Human Paleoecology and Foraging Theory in the Great Basin. In *Models for the Millennium: Great Basin Anthropology Today*, edited by C. Beck, pp. 141-151. University of Utah Press, Salt Lake City.

Green, D.G.
1981 Time Series and Postglacial Forest Ecology. *Quaternary Research* 15:265-277.

Greig, J.
1992 The Deforestation of London. *Review of Palaeobotany and Palynolo*gy 73:71-86.

Griffin, D.
2002 Prehistoric Human Impacts on Fire Regimes and Vegetation in the Northern Intermountain West. In *Native Peoples and the Natural Landscape*, edited by T.R. Vale, pp. 77-100. Island Press, Washington, D.C.

Grossman, E.L., and T.L. Ku
1986 Oxygen and Carbon Isotope Fractionation in Biogenic Aragonite: Temperature Effects. *Chemical Geology* 59:59-74.

Grove, J.M., and R. Switsur
1994 Glacial Geological Evidence for the Medieval Warm Period. *Climatic Change* 26:143-169.

Guyette, R.P., and D.C. Dey
2000 Humans, Topography, and Wildland Fire; The Ingredients for Long-term Patterns in Ecosystems. In *Proceedings: Workshop on Fire, People and the Central Hardwoods Landscape. March 12-14 2000 Richmond, Kentucky*. USDA Forest Service Northeastern Research Station General Technical Report NE-274, pp. 28-35

Guyette, R.P., R.M. Muzika, and D.C. Dey
2002 Dynamics of an Anthropogenic Fire Regime. *Ecosystems* 5:472-486

Guyette, R.P., and M.A. Spetich

2002 Fire History of Oak-Pine Forests in the Lower Boston Mountains, Arkansas, USA. *Forest Ecology and Management* 280:463-474

Guyette, R.P., M.A. Spetich, and M.C. Stambaugh
2006 Historic Fire Regime Dynamics and Forcing Factors in the Boston Mountains, Arkansas, USA. *Forest Ecology and Management* 234:293-304.

Hall, F.C.
1976 Fire and Vegetation in the Blue Mountains -- Implications for Land Managers. *Proceedings of the 15ᵗʰ Tall Timbers Fire Ecology Conference*, pp. 155-170.

Hannon, G.E., and R.H.W. Bradshaw
2000 Impacts and Timing of the First Human Settlement on Vegetation of the Faroe Islands. *Quaternary Research* 54:404-413.

Hamilton, J.
1994 *Time Series Analysis*. Princeton University Press, Princeton.

Hammet, J.E.
1992 Ethnohistory of Aboriginal Landscapes in the Southern United States. *Southern Indian Studies* 41:1-50.

Head, L.
1989 Prehistoric Aboriginal Impacts on Australian Vegetation: An Assessment of the Evidence. *Australian Geographer* 20(1):37–46.
2007 Cultural Ecology: The Problematic Human and the Terms of Engagement. *Progress in Human Geography* 31(6):837-846.

Heikens, A.L., and P.A. Robertson
1994 Barrens of the Midwest: A Review of the Literature. *Castanea* 3(59):184-194.

Heiri, O., A.F. Lotter, and G. Lemcke
2001 Loss on Ignition as a Method for Estimating Organic and Carbonate Content in Sediments: Reproducibility and Comparability of Results *Journal of Paleolimnology* 25(1):101-110.

Hemphill, M. L.
1983 *Fire, Vegetation, and People—Charcoal and Pollen Analysis of Sheep Mountain Bog, Montana, the Last 2800 Years*. Master's Thesis, Washington State University, Pullman.

Herzschuh, U., C. Zhang, S. Mischke, R. Herzschuh, F. Mohammadi, B. Mingram, H. Kürschner and F. Riedel
2005 A Late Quaternary Lake Record from the Qilian Mountains (NW China): Evolution of the Primary Production and the Water Depth Reconstructed from Macrofossil, Pollen, Biomarker, and Isotope Data. *Global and Planetary Change* 46(1-4):361-379.

Heusser, C.J., L.E. Heusser, and D.M. Peteet
1985   Late-Quaternary Climatic Change on the American North Pacific Coast. *Nature* 315(6019):485-487.

Heusser, C.J., L.E. Heusser, and S.S. Streeter
1980   Quaternary Temperatures and Precipitation for the North-west Coast of North America. *Nature* 286(5774):702-704.

Hicks, R.R., Jr.
2000   Humans and Fire: A History of the Central Hardwoods. In *Proceedings: Workshop on Fire, People and the Central Hardwoods Landscape. March 12-14 2000 Richmond, Kentucky*. USDA Forest Service Northeastern Research Station General Technical Report NE-274, pp. 3-18

Hoblitt, R.P., D.R. Crandell, D.R. Mullineaux
1980   Mount St. Helens Eruptive Behavior during the Past 1,500 Years. *Geology* 8:555-559.

Hodell, D.A., M. Brenner and J. Curtis
2005   Terminal Classic Drought in the Northern Maya Lowlands Inferred from Multiple Sediment Cores in Lake Chichancanab (Mexico). *Quaternary Science Reviews* 24: 1413–1427.

Horn, H.S.
1975   Forest Succession. *Scientific American* 232(5):90-98.

Hughes, M.K., and H.F. Diaz
1994   Was There a 'Medieval Warm Period', and if so, Where and When? *Climatic Change* 26(2-3):109-142.

Hunn, E.S.
1990   *Nch'i-Wána, "The Big River": Mid-Columbia Indians and Their Land*. University of Washington Press, Seattle.
1999   Mobility as a Factor Limiting Resource Use on the Columbia Plateau. In *Northwest Lands, Northwest Peoples*, edited by D.D. Goble and P.W. Hirt, pp. 156-172. University of Washington Press, Seattle.
2000   On the Relative Contributions of Men and Women to Subsistence of Hunter-Gatherers of the Columbia Plateau: A Comparison with *Ethnographic Atlas* Summaries. In *Ethnobotany: A Reader*, edited by P.E. Minnis, pp. 184-196. University of Oklahoma Press, Norman, OK.

Hunt, T.L., and C.P. Lipo
2006   Late Colonization of Easter Island. *Science* 311(5767):1603 - 1606

Iversen, J.
1956   Forest Clearance in the Stone Age. *Scientific American* 194(3):36-41.

Jacobson, G.L., Jr.
1979   The Palaeoecology of White Pine (*Pinus strobus*) in Minnesota. *Journal of Ecology* 67(2):697-726.

Jacobson, G.L., Jr., and R.H.W. Bradshaw
1981   The Selection of Sites for Paleovegetational Studies. *Quaternary Research* 16:80-96.

Jacobson, G.L., and E.C. Grimm
1986   A Numerical Analysis of Holocene Forest and Prairie Vegetation in Central Minnesota. *Ecology* 67:958-966.

Janssen, C.R.
1986   The Use of Local Pollen Indicators of the Contrast between Regional and Local Pollen Values in the Assessment of Human Impact on Vegetation. In *Anthropogenic Indicators in Pollen Diagrams*, edited by K.E. Behre, pp. 203-208. A.A. Balkema, Boston.

Jermann, J.V., and R.D. Mason
1976   *A Cultural Resource Overview of the Gifford Pinchot National Forest of South-Central Washington*. Office of Public Archaeology, Institute for Environmental Studies, University of Washington, Seattle.

Jones, R.
1969   Firestick Farming. *Australian Natural History* 16:224–231.

Jones, T.L., G.M. Brown, M. Raab, J.L. McVickar, W.G. Spaulding, D.J. Kennett, A. York, and P.L. Walker
1999   Environmental Imperatives Reconsidered: Demographic Crises in Western North America During the Medieval Climatic Anomaly. *Current Anthropology* 40(2):137-170.

Kapp, R.O.
1969   *How to Know Pollen and Spores*. W.C. Brown Co, Dubuque.

Kay, C.E.
1994   Aboriginal Overkill: The Role of Native Americans in Structuring Western Ecosystems. *Journal Human Nature* 5(4):359-398.
2000   Native Burning in Western North America: Implications for Hardwood Forest Management. In *Proceedings: Workshop on Fire, People, and the Central Hardwoods Landscape, General Technical Report NE-274*, edited by D.A. Yaussy, pp. 19-27. U.S. Department of Agriculture, Forest Service, Northeastern Research Station, Newtown Square.
2007   Are Lightning Fires Unnatural? A Comparison of Aboriginal and Lightning Ignition Rates in the United States. In *Proceedings of the 23rd Tall Timbers Fire Ecology Conference: Fire in Grassland and Shrubland Ecosystems*, edited by R.E. Masters and K.E.M. Galley, pp. 16-28. Tall Timbers Research Station, Tallahassee, FL.

Kealhofer, L. and B.J. Baker
1996 Counterpoint to Collapse: Depopulation and Adaptation. In *Bioarchaeology of Native American Adaptation in the Spanish Borderlands*, edited by B.J. Baker and L. Kealhofer, pp. 209-222. University Press of Florida, Gainesville.

Keeley, J.E.
2002 Native American Impacts on Fire Regimes of the California Coastal Ranges. *Journal of Biogeography* 29:303–320.

Kirch, P.V.
1982 The Impact of the Prehistoric Polynesians on the Hawaiian Ecosystem. *Pacific Science* 36(1):1-14.
1997 Microcosmic Histories: Island Perspectives on "Global" Change. *American Anthropologist* 99:30-42.

Kirch, P.V., and J. Ellison
1994 Palaeoenvironmental Evidence for Human Colonization of Remote Oceanic Islands. *Antiquity* 68:310-321.

Kirch, P.V., J.R. Flenley, D.W. Steadman, F. Lamont and S. Dawson
1992 Ancient Environmental Degradation, Prehistoric Human Impacts on an Island Ecosystem: Mangaia, Central Polynesia. *National Geographic Research & Exploration* 8(2):166-179.

Kirch, P.V. and Hunt, T.L.
1997 *Historical Ecology in the Pacific Islands: Prehistoric Environmental and Landscape Change*. Yale University Press, New Haven.

Kirk, R., and R.D. Daugherty
2007 *Archaeology in Washington*. University of Washington Press, Seattle.

Konrad, J.G., G. Chesters, and D.R. Keeney
1970 Determination of Organic- and Carbonate-carbon in Freshwater Lake Sediments by a Microcombustion Procedure. *Journal of Thermal Analysis and Calorimetry* 2:199-208.

Krech, S.
1999 *The Ecological Indian: Myth and History*. W.W. Norton & Co., New York.

Kruckeberg, A.R.
1991 *Natural History of the Puget Sound Country*. University of Washington Press, Seattle.
1999 A Natural History of the Puget Sound Basin. In *Northwest Lands, Northwest Peoples*, edited by D.D. Goble and P.W. Hirt, pp. 51-78. University of Washington Press, Seattle.

Lacourse, T.,R.W. Mathewes, and R.J. Hebda
2007 Paleoecological Analyses of Lake Sediments Reveal Prehistoric Human Impact on Forests at Anthony Island UNESCO World Heritage Site, Queen Charlotte Islands (Haida Gwaii), Canada. *Quaternary Research* 68:177-183.

Lahren, S.L., Jr.
1998 Kalispel. In *Plateau*, edited by D.E. Walker Jr., pp. 283-296. Handbook of North American Indians, Volume 12, W.C. Sturtevant, general editor, Smithsonian Institution, Washington D.C.

Langston, N.
1999 Human and Ecological Change in the Inland Northwest Forests. In *Northwest Lands, Northwest Peoples*, edited by D.D. Goble and P.W. Hirt, pp. 415-436. University of Washington Press, Seattle.

Leavell, D., and J.C. Chatters
1996 *Anthropogenic Fire: Applying the Artifact Concept*. Paper delivered to the Society for American Archaeology conference, New Orleans, LA. April 11, 1996.

Leeds, L.L., L.A. Leeds, and K.A. Whittlesey
1985 Model Building as an Approach to Explaining the Evolution of Hunter-Gatherer Adaptations on the Columbia Plateau. In *Summary Report, Chief Joseph Dam Cultural Resources Project*, edited by S.K. Campbell, pp. 3-81. Office of Public Archaeology, University of Washington, Seattle.

Leng, M., P. Barnker, P. Greenwood, N. Roberts, and J. Reed
2001 Oxygen Isotope Analysis of Diatom Silica and Authigenic Calcite from Lake Pinarbasi, Turkey. *Journal of Paleolimnology* 25(3):343-349.

Leopold, E.B. and R. Boyd
1999 An Ecological History of Old Prairie Areas in Southwestern Washington. In *Indians, Fire, and The Land in the Pacific Northwest*, edited by R. Boyd, pp. 139-163. Oregon State University Press, Corvallis.

Lewis, H.T.
1982 Fire Technology and Resource Management in Aboriginal North America and Australia. In *Resource Managers: North American and Australian Hunter-Gatherers*, edited by N.M. Williams and E.S. Hunn, pp. 45-68. Westview Press, Inc, Boulder.
1993 Patterns of Indian Burning in California: Ecology and Ethnohistory. In *Before the Wilderness: Environmental Management by Native Californians*, edited by T.C. Blackburn and K. Anderson, pp. 55-116. Ballena Press, Menlo Park, California. (Originally published in 1973 as Ballena Press Anthropological Papers 1).

Lewis, H.T., and T.A. Ferguson.
1988 Yards, Corridors, and Mosaics: How to Burn a Boreal Forest. *Human Ecology* 16:57-77.

Linse, A.R.
1993 Geoarchaeological Scale and Archaeological Interpretation: Examples from the Central Jornada Mogollon. In *Effects of Scale on Archaeological and Geoscientific Perspectives*, edited by J.K. Stein, J.K. and A.R. Linse, pp. 11-28. Geological Society of America Special Paper 283. Boulder, Co.

Lohse, E.S., and R. Sprague
1998 History of research. In *Plateau*, edited by D.E. Walker Jr., pp. 8-28. Handbook of North American Indians, Volume 12, W.C. Sturtevant, general editor, Smithsonian Institution, Washington D.C.

Luckman, B.H.
1994 Evidence for Climatic Conditions between ca. 900-1300 A.D. in the Southern Canadian Rockies. *Climatic Change* 26:171-182.

Mack, R.N.
1988 First Comprehensive Botanical Survey of the Columbia Plateau, Washington: The Sandberg and Leiberg Expedition of 1893. *Northwest Science* 62(3)118-128.

Mack, R.N., V.M. Bryant Jr., and W. Pell
1978 Modern Forest Pollen Spectra from Eastern Washington and Northern Idaho. *Botanical Gazette* 139:249-255.

Mack, R.N., V.M. Bryant Jr., and S. Valastro
1977 Late Pleistocene Pollen Sequence from Hager Pond, Bonner Co., Idaho. *Palynology* 1:175.
1976 Pollen Sequence from the Columbia Basin, Washington; Reappraisal of Postglacial Vegetation. *American Midland Naturalist* 95:390-397.

Mack, R.N., N.W. Rutter, and S. Valastro
1979 Holocene Vegetation History of the Okanogan Valley, Washington. *Quaternary Research* 12:212-225.
1983 Holocene Vegetational History of the Kootenai River Valley, Montana. *Quaternary Research* 20(2):177-193.

Mack, R.N., N.W. Rutter, S. Valastro, and V.M. Bryant Jr.
1978 Late Quaternary Vegetation History at Waits Lake, Colville River Valley, Washington. *Botanical Gazette* 139:499-506.

Mack, R.N., N.W. Rutter, V.M. Bryant Jr., and S. Valastro
1978 Late Quaternary Pollen from Big Meadow, Pend Oreille County, Washington. *Ecology* 59:956-966.

Macklin, M.G., C. Bonsall, F.M. Davies, and M.R. Robinson
2000 Human-Environment Interactions during the Holocene: New Data and Interpretations from the Oban Area, Argyll, Scotland. *The Holocene* 10:109-121.

Maher, L.J., Jr.
1972 Absolute Pollen Diagram of Redrock Lake, Boulder County, Colorado. *Quaternary Research* 2:531-553.
1981 Statistics for Microfossil Concentration Measurements Employing Samples Spiked with Marker Grains. *Review of Palaeobotany and Palynology* 32:153-191.

Mann, C.C.
2005 *1491: New Revelations of the Americas Before Columbus.* Alfred A. Knopf, New York.

Mann, R.B.
2000 Fire in the Central Hardwoods: Why Are We Concerned? In *Proceedings: Workshop on Fire, People and the Central Hardwoods Landscape. March 12-14 2000 Richmond, Kentucky.* USDA Forest Service Northeastern Research Station General Technical Report NE-274, pp 1-2.

Marlon, J., Bartlein, P.J., Whitlock, C.
2006 Fire-fuel-climate Linkages in the Northwestern USA during the Holocene. *The Holocene* 16(8):1059-1071.

Marshall, A.G.
1999 Unusual Gardens: The Nez Perce and Wild Horticulture on the Eastern Columbia Plateau. In *Northwest Lands, Northwest Peoples*, edited by D.D. Goble and P.W. Hirt, pp. 173-187. University of Washington Press, Seattle.

Martin, P.S.
1967 Prehistoric Overkill. In *Pleistocene Extinctions: The Search for a Cause*, edited by P.S. Martin and H.E. Wright Jr., pp. 75-120. Yale University Press, New Haven.
1972 The Discovery of America. *Science*: 179:969-974.
1984 Prehistoric Overkill: The Global Model. In *Quaternary Extinctions: A Prehistoric Revolution*, edited by P.S. Martin and R.G. Klein, pp. 354-403. University of Arizona Press, Tucson.

Martin, P.S., and D.W. Steadman
1999 Prehistoric Extinctions on Islands and Continents. In *Extinctions in Near Time*, edited by R.D.E. MacPhee, pp. 17-52. Kluwer, New York.

Martin, R.E., J.D. Dell, and L.P. Neuenschwander
1977 Planning for Prescribed Burning in the Inland Northwest. In *Region 6 Eastside Prescribed Burning Workshop, Okanogan National Forest*

and Colville Indian Reservation, October 3-7, 1977. USDA Forest Service, Seattle.

Maruoka, K.R.
1994 *Fire History of* Pseudotsuga menziesii *and* Abies grandis *stands in the Blue Mountains of Oregon and Washington.* Master's Thesis, University of Washington, Seattle, WA.

Mayewski, P.A., and E.E. Rohling, J.C. Stager, W. Karlén, K.A. Maasch, L.D. Meeker, E.A. Meyerson, F. Gasse, S. van Kreveld, K. Holmgren, J. Lee-Thorp, G. Rosqvist, F. Rack, M. Staubwasser, R.R. Schneider, and E.J. Steig
2004 Holocene Climate Variability. *Quaternary Research* 62:243-255.

McAndrews, J.H.
1976 Man's Impact on the Canadian Flora. *Canadian Botanical Association Bulletin*, Supplement to Vol. 9 No 1.
1988 Human Disturbance of North American Forests and Grasslands: The Fossil Pollen Record. In *Vegetation History*, edited by B. Huntley and T. Webb III, pp. 673-697. Kluwer Academic Publishers.

McBride, J.R.
1983 Analysis of Tree Rings and Fire Scars to Establish Fire History. *Tree-Ring Bulletin* 43:51-67.

McConnaughey, T.A.
1986 *Oxygen and Carbon Isotope Disequilibria in Galapagos Corals: Isotopic Thermometry and Calcification Physiology.* Ph.D. Dissertation, University of Washington, Seattle.

McDadi, O., and R.J. Hebda
2008 Change in Historic Fire Disturbance in a Garry Oak (*Quercus garryana*) Meadow and Douglas-fir (*Pseudotsuga menziesii*) Mosaic, University of Victoria, British Columbia, Canada: A Possible Link with First Nations and Europeans. *Forest Ecology and Management* 256:1704-1710.

McGlone, M.S.
1983 Polynesian Deforestation of New Zealand: A Preliminary Synthesis. *Archaeology in Oceania* 18:11-25
1989 Polynesian Settlement of New Zealand in Relation to Environmental and Biotic Changes. *New Zealand Journal of Ecology* 12:115-129.

McGlone, M.S., and J.M. Wilmshurst
1999 Dating Initial Maori Environmental Impact in New Zealand. *Quaternary International* 59:5-16.

Meese, D.A., A.J. Gow, P. Grootes, P.A. Mayewski, M Ram, M Stuiver, K.C. Tayor, E.D. Waddington, and G.A. Zielinski

1994 The Accumulation Record from the GISP2 Core as an Indicator of Climate Change throughout the Holocene. *Science* 266(5191):1680-1682.

Mehringer, P.J., Jr.
1985 Late-Quaternary Pollen Records from the Interior Pacific Northwest and Northern Great Basin of the United States. *In Pollen Records of Late-Quaternary North American Sediments*, edited by V.A. Bryant Jr. and R.G. Holloway, pp. 167-189. American Association of Stratigraphic Palynologists, Dallas.

Mehringer, P.J., Jr., S.F. Arno and K.L. Petersen
1977a Postglacial History of Lost Trail Pass Bog, Bitterroot Mountains, Montana. *Arctic and Alpine Research* 9(4):345-368.

Mehringer, P.J., Jr., E. Blinman, and K.L Petersen
1977b Pollen Influx and Volcanic Ash. *Science* 198(4314):257-261.

Miller, J.
1998 Middle Columbia River Salishans. In *Plateau*, edited by D.E. Walker Jr., pp. 253-270. Handbook of North American Indians, Volume 12, W.C. Sturtevant, general editor, Smithsonian Institution, Washington D.C.

Millspaugh, S.H., and C. Whitlock
1995 A 750-year Fire History Based on Lake Sediment Records in Central Yellowstone National Park, USA. *The Holocene* 5(3):283-292.

Millspaugh, S.H., C. Whitlock, and P.J. Bartlein
2000 Variations in Fire Frequency and Climate over the Past 17000 yr in Central Yellowstone National Park. *Geology* 28(3):211-214.

Mohr, J.A., C. Whitlock, and C.N. Skinner
2000 Postglacial Vegetation and Fire History, Eastern Klamath Mountains, California, USA. *The Holocene* 10(5):587-601.

Fulé, P.Z., W. Covington, and M.M. Moore
1997 Determining Reference Conditions for Ecosystem Management of Southwestern Ponderosa Pine Forests. *Ecological Applications* 7(3):895–908

Moore, P.D., J.A. Webb, and M.E. Collins
1991 *Pollen Analysis.* Blackwell Scientific Publishing, Oxford, Boston.

Moreno, P.I.
2000 Climate, Fire, and Vegetation between about 13,000 and 9200 ^{14}C year B.P. in the Chilean Lake District. *Quaternary Research* 54(1):81-89.

Morrison, K.D.
1994 Monitoring Regional Fire History through Size-

specific Analysis of Microscopic Charcoal: The Last 600 Years in South India. *Journal of Archaeological Science* 21(5):675-685.

Nagaoka, L.
2000 Resource Depression, Extinction, and Subsistence Change in Southern New Zealand. Ph.D. Dissertation, University of Washington, Seattle
2001 Using Diversity Indices to Measure Changes in Prey Choice at the Shag River Mouth Site, Southern New Zealand. *International Journal of Osteoarchaeology* 11:101-111.

Naroll, R.
1962 Floor Area and Settlement Population. *American Antiquity* 27(4):587-589.

Nesje, A. and S.O. Dahl
2001 The Greenland 8200 cal. yr BP Event Detected in Loss-on-ignition Profiles in Norwegian Lacustrine Sediment Sequences. *Journal of Quaternary Science* 16(2):155-166.

Nickmann, R.J., and E. Leopold
1985 A Postglacial Pollen Record from Goose Lake, Okanogan County, Washington: Evidence for an Early Holocene Cooling. In *Summary Report, Chief Joseph Dam Cultural Resources Project*, edited by S.K. Campbell, pp. 131-147. Office of Public Archaeology, University of Washington, Seattle.

Noon, P.E., M.J. Leng and V.J. Jones
2003 Oxygen-Isotope (δ^{18}O) Evidence of Holocene Hydrological Changes at Signy Island, Maritime Antarctica. *The Holocene* 13(.2):251-263.

NOAA
2001 *Review of the Archaeological Data* (Chapter 2 in the Cultural Affiliation Report). Data retrieved from the National Park Service web site on Kennewick Man: http://www.cr.nps.gov/aad/kennewick/ames.htm

Ogden, J., L. Basher, and M. McGlone
1998 Fire, Forest Regeneration and Links with Early Human Habitation: Evidence from New Zealand. *Annals of Botany* 81(6):687-696.

Ostrom, C.W., Jr.
1990 *Time Series Analysis: Regression Techniques.* Sage Publications, Newbury Park.

Palmer, G.B.
1998 Coeur d'Alene. In *Plateau*, edited by D.E. Walker Jr., pp. 313-326. Handbook of North American Indians, Volume 12, W.C. Sturtevant, general editor, Smithsonian Institution, Washington D.C.

Parker, K.C.

2002a Fire in the Pre-European Lowlands of the American Southwest. In *Native Peoples and the Natural Landscape*, edited by T.R. Vale, pp. 101-141. Island Press, Washington, D.C.
2002b Fire in Sierra Nevada Forests: Evaluating the Ecological Impact of Burning by Native Americans. In *Native Peoples and the Natural Landscape*, edited by T.R. Vale, pp. 233-267. Island Press, Washington, D.C.

Patterson, W.A., III, and A.E. Backman
1988 Fire and Disease History of Forests. In *Vegetation History*, edited by B. Huntley and T. Webb III, pp. 603-632. Kluwer Academic Publishers, New York.

Patterson, W.A., III, K.J. Edwards, D.G. MacGuire
1987 Microscopic Charcoal as a Fossil Indicator of Fire. *Quaternary Science Reviews* 6:3-23.

Patterson W.A., III, and K.E. Sassaman.
1988 Indian Fires in the Prehistory of New England. In *Holocene Human Ecology in Northeastern North America*, edited by G.P. Nicholas, pp. 107-135. Plenum, New York.

Perttula, T.K.
1992 *"The Caddo Nation": Archaeological and Ethnohistoric Perspectives.* University of Texas Press, Austin.

Peter, D.H., and D. Shebitz
2006 Historic Anthropogenically Maintained Bear Grass Savannas of the Southeastern Olympic Peninsula. *Restoration Ecology* 14(4):605–615.

Plog, F.
1975 Demographic Studies in Southwestern Prehistory. *American Antiquity*, Memoir No. 30, 40(2)(Part 2):94-103.

Pyne, S. J.
1982 *Fire in America: A Cultural History of Wildland and Rural Fire.* Princeton University Press, Princeton, New Jersey.
1991 *Burning Bush: A Fire History of Australia.* Henry Holt and Company, New York.
1997 *Vestal Fire: An Environmental History.* University of Washington Press, Seattle.

Ramenofsky, A.F.
1987 *Vectors of Death: The Archaeology of European Contact.* University of New Mexico Press, Albuquerque.

Rauw, D.M.
1980 *Interpreting the Natural Role of Fire. Implications for Fire Management Policy.* Master's Thesis, University of Washington, Seattle, WA.

Ray, V.F.
1932 *The Sanpoil and Nespelem: Salishan Peoples of*

Northeast Washington. University of Washington, Publications in Anthropology 5.

Redman, C.L.
1999 *Human Impact on Ancient Environments.* The University of Arizona Press, Tucson, Arizona.

Redman, C.L., James, S.R., Fish, P.R., Rogers, J.D.,
2004 Introduction: Human Impacts on Past Environments. In *The Archaeology of Global Change: The Impact of Humans on their Environment*, edited by C.L. Redman, S.R. James, P.R. Fish, P.R. and J.D. Rogers, pp. 1-8. Smithsonian Books, Washington, D.C.

Rice, D.S.
1996 Paleolimnological Analysis in the Central Petén, Guatemala. In *The Managed Mosaic: Ancient Maya Agricultural and Resource Use*, edited by S.L. Fedick, pp. 193-206. University of Utah Press, Salt Lake City, Utah.

Rice, D.S, and P.M. Rice
1984 Lessons from the Maya. *Latin American Research Review* 19(3):7-34.

Rice, D.S., P.M. Rice, and E.S. Deevey Jr.
1985 Paradise Lost: Classic Maya Impact on a Lacustrine Environment. In *Prehistoric Lowland Maya Environment and Subsistence Economy*, edited by M. Pohl, pp. 91-105. Peabody Museum Papers, Vol. 77, Cambridge.

Robbins, W.G.
1999 Landscape and Environment: Ecological Change in the Intermontane Northwest. In *Indians, Fire, and The Land in the Pacific Northwest*, edited by R. Boyd, pp. 219-237. Oregon State University Press, Corvallis.

Romme, W.H., and D.G. Despain
1989 Historical Perspectives on the Yellowstone Fires of 1988. *Bioscience* 39(19):695-699.

Sabin, A.L., and N.G. Pisias
1996 Sea Surface Temperature Changes in the Northeastern Pacific Ocean during the Past 20,000 Years and Their Relationship to Climate Change in Northwestern North America. *Quaternary Research* 46(1):48-61.

Sarmaja-Korjonen, K., A. Seppänen, and O. Bennike
2006 *Pediastrum* Algae from the Classic Late Glacial Bølling Sø Site, Denmark: Response of Aquatic Biota to Climate Change. *Review of Palaeobotany and Palynology*, 138(2):95-107.

Sarna-Wojcicki, A.M., and O.K. Davis
1991 Quaternary Tephrochronology. In *The Geology of North America*, edited by R.B. Morrison, pp. 93-116. Vol. K-2, *Quaternary Nonglacial Geology: Coterminous U.S.* The Geological Society of America.

Sarna-Wojcicki, A.M., K.R. Lajoie, C.E. Meyer, and C.P. Adam
1991 Tephrochronologic Correlation of Upper Neogene Sediments along the Pacific Margin, Coterminous United States. In *The Geology of North America*, edited by R.B. Morrison, pp. 117-140. Vol. K-2, *Quaternary Nonglacial Geology: Coterminous U.S.* The Geological Society of America.

Schacht, R.M.
1981 Estimating Past Population Trends. *Annual Review of Anthropology* 10:119-140.

Scharf, E.A.
2002 *Long-term Interactions of Climate, Vegetation, Humans, and Fire in Eastern Washington.* Ph.D. Dissertation, University of Washington, Seattle, WA.

Schlegel, M.E., A.L. Mayo, , S. Nelson, D., Tingey, R. Henderson and D. Eggett
2009 Paleo-climate of the Boise Area, Idaho from the Last Glacial Maximum to the Present Based on Groundwater δ2H and δ18O Compositions. *Quaternary Research* 71(2):172-180.

Schoonmaker, P.K. and D.R. Foster
1991 Some Implications of Paleoecology for Contemporary Ecology. *The Botanical Review* 57(3):204-245.

Schwartz, D.W.
1956 Demographic Changes in the Early Periods of Cohonina Prehistory. In *Prehistoric Settlement Patterns in the New World*, edited by G.R. Wiley. Viking Fund Publications in Anthropology, No. 23.

Shackleton, N.J. and N.D. Opdyke
1973 Oxygen Isotope and Paleomagnetic Stratigraphy of Equatorial Pacific Core V28-238: Oxygen Isotope Temperatures and Ice Volumes on a 10^5-year Time Scale. *Quaternary Research* 3(1):39-55.

Shawcross, W.
1967 An Investigation of Prehistoric Diet and Economy on a Coastal Site at Galatea Bay, New Zealand. *Proceedings of the Prehistoric Society* 33:107-131.
1972 Energy and Ecology: Thermodynamic Models in Archaeology. In *Models in Archaeology*, edited by D.L. Clarke, pp. 577-622. Methuen Press, London.

Singh, G.
1980 A Long-term Pleistocene Vegetation, Climatic and Fire History Record from Southeastern New South Wales, Australia. *Abstracts—International Palynological Conference* 5:363.

Skinner, C., and A.G. Brown
1999 Mid-Holocene Vegetation Diversity in Eastern Cumbria. *Journal of Biogeography* 26(1):45-54.

Smith, C.S.
1983 *A 4300-year History of Vegetation, Climate, and Fire from Blue Lake, Nez Perce County, Idaho.* Master's Thesis, Washington State University, Pullman.

Smith, D., and C.P. Laroque
1996 Dendroglaciological Dating of a Little Ice Age Glacial Advance at Moving Glacier, Vancouver Island, British Columbia. *Géographie Physique et Quaternaire* 50(1):47-55.

Smith, H.W., R. Okazaki, and C. Knowles
1979 Electron Microprobe Analysis of Glass Shards from Tephra Assigned to Set W, Mount St. Helens, Washington. *Quaternary Research* 7:207-217.

Soeriaatmadja, R.E.
1966 *Fire History of the Ponderosa Pine Forests of Warm Springs Indian Reservation, Oregon.* Ph.D. Dissertation, Oregon State University, Corvallis.

Spier, L.
1936 *Tribal Distribution in Washington.* General Series in Anthropology 3.

Steadman, D.W.
1995 Prehistoric Extinctions of Pacific Island Birds: Biodiversity Meets Zooarchaeology. *Science* 267:1123-1131.

Steadman, D.W. and P.V. Kirch
1990 Prehistoric Extinction of Birds on Mangaia, Cook Island, Polynesia. *Proceedings of the National Academy of Sciences* 87(24):9605-9609.

Stein, J.K.
1993 Scale in Archaeology, Geosciences, and Geoarchaeology. In *Effects of Scale on Archaeological and Geoscientific Perspectives,* edited by J.K. Stein, J.K. and A.R. Linse, pp. 1-10. Geological Society of America Special Paper 283. Boulder, Co.

Stenholm, N.A.
1985 Botanical Assemblage. In *Summary Report, Chief Joseph Dam Cultural Resources Project,* edited by S.K. Campbell, pp. 421-453. Office of Public Archaeology, University of Washington, Seattle.

Stewart, O.C.
1956 Fire as the First Great Force Applied by Man. In *Man's Role in the Changing Face of the Earth,* edited by W.L. Thomas, pp. 115-133, University of Chicago Press, Chicago.

Stuiver, M.
1970 Oxygen and Carbon Isotope Ratios of Fresh-Water Carbonates as Climatic Indicators. *Journal of Geophysical Research* 75(27):5247-5257.

Stuiver, M., and P.J. Reimer
1993 Extended ^{14}C data base and revised CALIB 3.0 ^{14}C age calibration program. *Radiocarbon* 35:215:230.
2000 CALIB 4.3 radiocarbon calibration program. Quaternary Isotope Lab, University of Washington.

Sugita, S.
1993 A Model of Pollen Source Area for an Entire Lake Surface. *Quaternary Research* 39(2):239-244.

Sugita, S., and M. Tsukada
1982 The Vegetation History in Western North America. I. Mineral and Hall Lakes. *Japanese Journal of Ecology* 32(4):499-515.

Swain, A.M.
1973 A History of Fire and Vegetation in Northeastern Minnesota as Recorded in Lake Sediments. *Quaternary Research* 3:383-396.
1978 Environmental Changes during the Past 2000 Years in North-central Wisconsin: Analysis of Pollen, Charcoal, and Seeds from Varved Lake Sediments. *Quaternary Research* 10:55-68.

Swetnam, T.W., C.D. Allen and J.L. Betancourt
1999 Applied Historical Ecology: Using the Past to Manage the Future. *Ecological Applications* 9(4):1189-1206

Taruntani, T., R.N. Clayton, and T.K. Mayeda
1969 Effect of Polymorphism and Magnesium Substitution on Oxygen Isotope Fractionation Between Calcium Carbonate and Water. *Geochimica et Cosmochimica Acta* 33:987-996.

Tauber, H.
1965 Differential Pollen Dispersion and the Interpretation of Pollen Diagrams. *Danmarks Geologiske Undersogelse* Series II 89.

Taylor, A.H., M.B. Jordan, and J.A. Stephens
1998 Gulf Stream shifts following ENSO events. *Nature* 393(6686):638.

Taylor, R.E.
1987 *Radiocarbon Dating: An Archaeological Perspective.* Academic Press, Inc., New York.

Tevesz, M.J.S., E. Barrera, and S.F. Schweigien
1996 Seasonal Variation in Oxygen Isotopic Composition of Two Freshwater Bivalves: *Sphaerium striatinum* and *Anodonta grandis.* *Journal of Great Lakes Research* 22(4):906-916

Thornton, R.
1997   Aboriginal North American Population and
        Rates of Decline, ca. A.D. 1500-1900. *Current
        Anthropology* 38(2):310-315.

Timbrook, J., J.R. Johnson, and D.D. Earle
1982   Vegetation Burning by the Chumash. *Journal of
        California and Great Basin Anthropology*,
        4(2):163-186.

Tsukada, M., S. Sugita, and Y. Tsukada
1986   Oldest Primitive Agriculture and Vegetational
        Environments in Japan. *Nature* 322(6080):632-
        634.

Turner, C.G., and L. Lofgren
1966   Household Size of Prehistoric Western Pueblo
        Indians. *Southwestern Journal of Archaeology*
        22(1):117-132.

Turner, N.J.
1999   "Time to Burn". In *Indians, Fire, and The Land
        in the Pacific Northwest*, edited by R. Boyd, pp.
        185-218. Oregon State University Press,
        Corvallis.

Umbanhower, C.E., Jr.
1996   Recent Fire History of the Northern Great
        Plains. *American Midland Naturalist* 135:115-
        121.

Vale, T.R.
2002a   The Pre-European Landscape of the United
        States: Pristine or Humanized? In *Native
        Peoples and the Natural Landscape*, edited by
        T.R. Vale, pp. 1-39. Island Press, Washington,
        D.C.
2002b   Reflections. In *Native Peoples and the Natural
        Landscape*, edited by T.R. Vale, pp. 295-301.
        Island Press, Washington, D.C.

Walker, D.E., Jr.
1998   Introduction. In *Plateau*, edited by D.E. Walker
        Jr., pp. 1-7. Handbook of North American
        Indians, Volume 12, W.C. Sturtevant, general
        editor, Smithsonian Institution, Washington
        D.C.

Weaver, H.
1959   Ecological Changes in the Ponderosa Pine
        Forest of the Warm Springs Indian Reservation
        in Oregon. *Journal of Forestry* 57(1):15-20.
1961   Ecological Changes in the Ponderosa Pine
        Forest of Cedar River Valley in Southern
        Washington. *Ecology* 42(2):416-420.
1967   Some Effects of Prescribed Burning on the
        Coyote Creek Test Area. *Journal of Forestry*
        65:552-558.

Webb, R.S., and T. Webb III
1988   Rates of Sediment Accumulation in Pollen Cores
        from Small Lakes and Mires of Eastern North
        America. *Quaternary Research* 30:284-297.

Weddell, B.J.
2001   Changing Perspectives in Nineteenth Century
        Written Descriptions of Palouse and Canyon
        Grasslands. *Idaho Bureau of Land Management.
        Technical Bulletin No 01-13. August 2001.*
        BLM, Cottonwood, Idaho.

Weng, C.
2005   An Improved Method for Quantifying
        Sedimentary Charcoal via a Volume Proxy. *The
        Holocene* 15:298-301.

Westgate, J.A. and M.P. Gorton
1981   Correlation Techniques in Tephra Studies. In
        Tephra Studies, edited by S. Self and R.S.J.
        Sparks, pp. 73-94. D. Reidel Publishing
        Company, Boston.

White, R.
1999   Indian Land Use and Environmental Change. In
        *Indians, Fire, and The Land in the Pacific
        Northwest*, edited by R. Boyd, pp. 36-49.
        Oregon State University Press, Corvallis.

Whitehead, D.R, and M.C. Sheehan
1985   Holocene Vegetational Changes in the
        Tombigbee River Valley. *Eastern Mississippi
        American Midland Naturalist* 113 (1):122-137.

Whitlock, C., and Knox, M.A.
2002   Prehistoric Burning in the Pacific Northwest:
        Human Versus Climatic Influences. In *Native
        Peoples and the Natural Landscape*, edited by
        T.R. Vale, pp. 195-231. Island Press,
        Washington, D.C.

Whitlock, C., and C.P.S. Larsen
2001   Charcoal as a Fire Proxy. *In Tracking
        Environmental Change using Lake Sediments.
        Volume 3: Terrestrial, Algal, and Siliceous
        Indicators*, edited by J.P. Smol, H.J.B Birks, and
        W.M., Last, pp. 75-97. Kluwer Academic
        Publishers, Dordrechet, pp. 75-97.

Whitlock, C., and P.J. Bartlein
1997   Vegetation and Climate Change in Northwest
        America during the Past 125 kyr. *Nature*
        388(6637):57-61.

Whitlock, C., and S.H. Millspaugh
1996   Testing the Assumptions of Fire-history Studies:
        An Examination of Modern Charcoal
        Accumulation in Yellowstone National Park,
        USA. *The Holocene* 6(1):7-15.

Winkler, M.G.
1985a   A 12,000-Year History of Vegetation and
        Climate for Cape Cod, Massachusetts.
        Quaternary Research 23:301-312.
1985b   Charcoal Analysis for Paleoenvironmental
        Interpretation: A Chemical Assay. *Quaternary
        Research* 23:313-326.

Wray, J. and M.K. Anderson
  2003   Restoring Indian-Set Fires to Prairie Ecosystems on the Olympic Peninsula. *Ecological Restoration* 21(4):296-301.

Wright, H.E., Jr., and H.L. Patten
  1963   The pollen sum. *Pollen et Spores* 5:445-450.

Wright, H.E., Jr., J.E. Kutzbach, and T. Webb III
  1993   *Global Climates since the Last Glacial Maximum.* University of Minneapolis Press, Minneapolis.

Wu, J., G.H. Schleser, A. Locke and S. Li
  2007   A Stable Isotope Record from Freshwater Lake Shells of the Eastern Tibetan Plateau, China, During the Past Two Centuries. *Boreas* 36:38-46.

Yamaguchi
  1983   New Tree-ring Dates for Recent Eruptions of Mt. St. Helens. *Quaternary Research* 10:246-250.
  1985   Tree-ring Evidence for a Two-year Interval between Recent Prehistoric Eruptions of Mt. St. Helens. *Geology* 13(8):554-557.

www.ingramcontent.com/pod-product-compliance
Lightning Source LLC
Chambersburg PA
CBHW061000030426
42334CB00033B/3299